ROUTLEDGE LIBR/
POLITICAL GE(

Volume 4

MONEY AND VOTES

MONEY AND VOTES
Constituency Campaign Spending and Election Results

R.J. JOHNSTON

LONDON AND NEW YORK

First published in 1987

This edition first published in 2015
by Routledge
2 Park Square, Milton Park, Abingdon, Oxon, OX14 4RN

and by Routledge
711 Third Avenue, New York, NY 10017

Routledge is an imprint of the Taylor & Francis Group, an informa business

British Library Cataloguing in Publication Data
A catalogue record for this book is available from the British Library

ISBN: 978-1-138-80830-0 (Set)
eISBN: 978-1-315-74725-5 (Set)
ISBN: 978-1-138-79991-2 (Volume 4)
eISBN: 978-1-315-75579-3 (Volume 4)
Pb ISBN: 978-1-138-79993-6 (Volume 4)

Publisher's Note
The publisher has gone to great lengths to ensure the quality of this reprint but
points out that some imperfections in the original copies may be apparent.

Disclaimer
The publisher has made every effort to trace copyright holders and would
welcome correspondence from those they have been unable to trace.

Printed and bound by CPI Group (UK) Ltd, Croydon, CR0 4YY

MONEY AND VOTES:

Constituency Campaign Spending and Election Results

R.J. JOHNSTON

CROOM HELM
London ● New York ● Sydney

© 1987 R.J. Johnston
Croom Helm Ltd, Provident House,
Burrell Row, Beckenham, Kent BR3 1AT
Croom Helm Australia, 44–50 Waterloo Road,
North Ryde, 2113, New South Wales

Published in the USA by
Croom Helm
in association with Methuen, Inc.
29 West 35th Street,
New York, NY 10001

British Library Cataloguing in Publication Data
Johnston, R.J.
    Constituency campaign spending and election
    results : analyses of post-war trends in
    the United Kingdom. — (Croom Helm series
    in geography and environment).
    1. Great Britain. *Parliament* — Elections
    — History — 20th century 2. Elections —
    Great Britain — Campaign funds — History —
    20th century
    I. Title
    324.7'8'0942   JN1039
    ISBN 0–7099–1466–0

Library of Congress Cataloging in Publication Data

ISBN 0–7099–1466–0

Photosetting by Mayhew Typesetting, Bristol, England
Printed and bound in Great Britain
by Billing & Sons Limited, Worcester.

Contents

Figures

Tables

Preface

Writing in *The Times* in 1980 (10 March), Michael Pinto-Duschinsky asked 'What should be the cost of a vote?' This was a contribution to what was then quite a lively political debate regarding the financing of political parties and the desirability of providing them with state aid for both their routine and their campaigning activities. A government-sponsored committee of enquiry had reported in 1976 that such state aid should be introduced, though this was never acted upon.

Most British political scientists who consider the issue are not in favour of state aid, for a variety of reasons. In his article in *The Times*, Pinto-Duschinsky argued that the importance of the spending by parties is 'easily exaggerated': thus, 'Given the limited sums spent on political publicity and the prohibition against paid advertising on radio and television, it is unlikely that the outcome of a general election ever has been or will be determined by the hidden persuaders'. The problem is how to analyse the impact of spending, so that such statements have a valid empirical base.

Most concern expressed on this topic refers to spending by the central party organisations, and especially their increased use of major advertising agencies. However, as Pinto-Duschinsky's data for the 1979 general election showed, as well as the national campaign (which is unrestricted apart from the ban on radio and TV advertising) which cost about £4m, a further large sum of about £3m was spent in the 635 constituencies. This spending was subject to strict legal regulation regarding amount, use, and reporting, and limits on the amounts have been increasingly restrictive since 1983. Pinto-Duschinsky thus wrote that

> Stringent legal restrictions on spending by parliamentary candidates (first introduced in 1983) have ensured approximate equality between the main parties and have greatly reduced the cost of constituency campaigning. In real terms, candidates in 1979 spent barely a quarter as much as in 1945.

The last statement is true, but the implication that what is spent has no impact is untested; the earlier statement regarding 'approximate equality' has become increasingly less valid since 1969. So a lot of money is spent in the constituency campaigns. What impact does it have?

The present book takes issue with the belief of Pinto-Duschinsky and others (notably the authors of the justly renowned Nuffield College reviews of British general elections) that constituency campaign expenditure is irrelevant. The work presented here developed from a series of small pieces of research on campaign spending in the constituencies — influenced by American writings and their much clearer modelling and analysis — which sought to provide a more detailed analysis of the British situation than had been published heretofore. A variety of models was tried, and eventually a satisfactory outcome was achieved in studies of the 1983 general election. This book uses that model for the period 1950–83.

I am indebted to a number of institutions and individuals for their contributions to the work reported here. Support for collecting the data and preparing them for analysis was provided by a grant from the University of Sheffield Research Fund; Mrs Barbara Woods did the work, and I am grateful to her for her care and efficiency. A grant from the Leverhulme Trust allowed me to employ Dr Araya Redda for four months as a research assistant. During that time, he operationalised all of the models and ran most of the analyses on the University of Sheffield computers; I am greatly in his debt for his ready appreciation of the tasks involved and for his conscientious work. The grant also covered the costs of producing this book, and I am grateful once again to Joan Dunn for her contribution to my research output. As the research developed over the last decade, Peter Taylor and Alan Hay have made their usual contributions in discussions of political and technical issues, and I am again in debt to them for their intellectual support. Finally, Rita Johnston has once more endured my preoccupation with a piece of research, though she too realised it was a welcome release, when possible, from the chores of committee chairmanship.

Saturday

. . . Spent the day helping at our Conservative fête and the sun shone on the godly, there must be easier ways for 35 people to raise £600, surely we must be worth more that £3 an hour? It might be a topic for a thesis, 'An analysis of fund raising activities in the Conservative Party and their relationship to electoral success.' Do those constituencies which specialize in jumble sales do better than those who are keen on whist drives? . . .

Tony Arbour, Don's Diary, *The Times Higher Educational Supplement*, 8.8.86, p. 4.

1

The Study of Voting

Elections of all kinds, but especially general elections, are the source of much discussion, speculation and analysis. They are, of course, crucial elements in the unfolding history of a country, since they are concerned with the allocation of power — ultimately to certain individuals but explicitly, in most cases, to political parties. Thus, the study of elections is a central element of the discipline of political science in liberal democracies. For geographers, elections are of interest because of their spatial components. Taylor (1978) has identified three separate, though linked, aspects of the study of elections which are inherently geographical (see also Johnston, 1980b; Taylor, 1985b): the geography of the inputs, which is concerned with the distribution of voters of various types and persuasions and of political parties and their activities; the geography of throughputs, which investigates the transformation of the geography of votes into a geography of representation, via the electoral law; and the geography of outputs, which relates to the allocation of political power and its consequences.

Unfortunately, there has been little joint development and application of the perspectives of geographers and political scientists to the study of elections in Britain (see Johnston and Taylor, 1986), and much more importantly, political scientists (who not surprisingly dominate the study of British elections) have in general paid little attention to the geographical perspective. As emphasised in a detailed review of their recent writings (Johnston, 1986b), most researchers note the presence of spatial variations in voting patterns (i.e. geography of inputs) and its impact on the geography of representation, but they seek neither to explain that phenomenon nor to incorporate it into their analytical frameworks for the study of voter behaviour.

1

The present book is written in the firmly held belief that the geographical perspective must be fully incorporated within British electoral studies. Its focus is on the geography of inputs, on the spatial variability in party activity during general election campaigns, and its goal is to evaluate hypotheses relating that activity to the electoral outcome. Voting behaviour in Britain cannot be appreciated, it is argued, solely as the outcome of people of particularly defined socio-economic types reacting to the information purveyed to them about and by the political parties through the mass media. They make their decisions in places, having been socialised in places, and the contexts provided by those milieux are not necessarily — indeed almost certainly are not — spatially invariant. To appreciate fully how voting behaviour is influenced, attention must be paid to the local as well as to the national contexts.

The present book is concerned with one aspect of the local context only — campaign activity by the political parties in the various constituencies, as indexed by campaign expenditure. To set that aspect of electoral study into context, however, the present chapter provides a brief introduction to British psephology before turning to a more detailed discussion of British political finance.

BRITISH PSEPHOLOGY: A BRIEF REVIEW

Elections produce a mass of data that are freely available for analysis. Most of these data refer to the election results — basically, the number of votes cast for each candidate. In the British first-past-the-post electoral system, the unit for reporting the number of votes cast is the constituency, because this is the spatial territory employed in the transformation of votes into seats; unlike some other countries which use the same electoral system (e.g. Australia, New Zealand, and the USA), no data are provided on the number of votes cast in subdivisions of the constituencies.

With over 600 constituencies in the United Kingdom, each general election produces very considerable amounts of information about the electoral geography of the country, and these data can be analysed in a variety of ways. Such analyses are typified by the Appendices to the Nuffield studies of every British general election since 1945 (the most recent being Butler and Kavanagh, 1984), which have been undertaken in recent years by Michael Steed and John Curtice. These studies rely almost entirely on the simple voting statistics (i.e. the percentage of the votes cast which were won by each party in each constituency), in part because of the goal of producing the book within a year of the election and in part because it is targeted at a wider audience than professional psephologists alone. Some analyses of particular geographical variations are attempted, however, looking, for example, at differences in party performance by region, by urban/rural location, and by level of unemployment in the constituencies. Recent elections have indicated a growing spatial polarisation of the electorate, described by Curtice and Steed (1984, p. 342) as follows:

> More depressed areas, principally the cities and the periphery, have been steadily moving towards Labour for a quarter of a century while more affluent areas, especially those of the South-East and the very rural seats, have moved in the opposite direction. The 1979–83 movements widened that division faster than ever before and added some interesting variations around the main cleavage, notably in delineating much more sharply the depressed inner areas of the five biggest conurbations.

To use these constituency electoral data in an analysis of voting behaviour, particularly in attempting to suggest reasons why people voted the way that they did (and where), it is desirable to know something of the people who live and vote in the constituencies. Simple regional and other aggregations can provide very useful pointers to the links between voting and socio-demographic characteristics (as in Taylor, 1979), and constituency vignettes (as in Crewe and Fox, 1984 and Waller, 1984) similarly facilitate useful qualitative insights. Detailed quantitative data on British constituencies have been made available from the population censuses since 1971 (though Miller, 1977, has produced a valuable data set for the period since 1921 by amalgamating constituencies so that they coincide with census limits), and these have provided the basic material for ecological studies (as reviewed in Taylor and Johnston, 1979; and Johnston, 1985a).

These ecological studies use the census data as independent variables and the electoral returns as dependent variables, exploring the links between the two and using the output to infer how different socio-demographic groups vote. Most such studies have concluded that groups vary from place to place in how they vote (as in Crewe and Payne, 1976); indeed Miller (1977, p. xiii) concluded that 'How people voted depended significantly more on where they lived than on their occupations', which he reinforced with his study of the 1983 results (Miller, 1984). These are, of course, only inferences from aggregate data. That such variations not only exist but are very substantial was confirmed in another study of the 1983 election (Johnston, 1985a) which used entropy-maximising procedures to estimate how people in particular groups voted in different places; at the constituency scale, the percentage of white-collar owner-occupiers who voted Conservative was estimated as varying between 20 and 58, for example, and the percentage of blue-collar renters who voted Labour as varying between 4 and 57.

These estimates use not only census and electoral data at the constituency scale but also national data (obtained from sample surveys) showing the link between voting and such attributes as occupation and housing tenure. The latter source is by far and away the most important for most psephologists today, because it avoids the ecological problem of inference that is present in the analysis of constituency data and provides a much greater range of material that can be linked to voting behaviour. Such surveys are usually small (most are based on about 2000 respondents; the 1983 British Election Study had 3955) which precludes much detailed analysis of spatial variations. This leads to the unfortunate implicit assumption underlying much British psephological writing that there is a uniform political culture that is spatially invariant (see Johnston, 1985a, 1986b) — an assumption that much of the geographical research has shown to be unfounded. (If it is unfounded, then the randomised sampling procedures which are the basis of the surveys are probably invalid: Miller, 1984; Warde, 1986.)

British voting behaviour as modelled in the British Election Studies

Of the various analyses of voting behaviour in Britain that are based on survey data, the most important are those in the sequence known as the British Election Studies: these began at the University of Oxford under David Butler's direction in the 1960s; were moved to the University of Essex in the 1970s, where they were directed by Ivor Crewe and B. Sarlvik; and returned to Oxford for the 1983 election with a study directed by Antony Heath, Roger Jowell and John Curtice. Their main output is a series of three books — *Political change in Britain* (Butler and Stokes, 1969 and 1974), *Decade of dealignment* (Sarlvik and Crewe, 1983), and *How Britain votes* (Heath, Jowell, and Curtice, 1985) — which present a common implicit model of British voting behaviour.

This model (whose characteristics are clearly delineated in the structure of the three books) has two basic components. The first refers to the long-term processes of *political socialisation*. To most analysts, the principal determinant of how people vote is their class position; they are socialised into a particular set of attitudes as a function of their occupation, and these are linked to one particular party. Thus, the Labour party has developed as the party of the working class, those with manual occupations; the Conservative party as the political voice of the middle class, those with non-manual occupations; and the Liberal party has not been associated with a particular set of attitudes but rather has been represented as a focus of protest votes only (Himmelweit *et al.*, 1985), though recent analysis suggests that a core Liberal/SDP Alliance ideology may now be developing (Heath, Jowell and Curtice, 1985). The concept of class and how it should be measured is a cause of much controversy (see Dunleavy and Husbands, 1985; Franklin, 1985; Kavanagh, 1986; Rose and McAlister, 1986).

Although primacy is given to the class cleavage as represented by occupation, other socio-economic charcteristics are also introduced as relevant to the processes of political socialisation. Parents' background is clearly important, since this provides the context for initial socialisation. Membership of a trade union is seen as enhancing the probability of voting Labour, given the links between the party and the trade-union movement. Housing tenure has been accorded increasing importance by some observers, as the Conservative party becomes clearly aligned with the private-market provision of major consumption items, including housing, whereas Labour is linked to a much greater level of social provision (of health care and education as well: see Dunleavy, 1979). Level of education is also now linked to voting; in particular, the better educated, especially in the salariat, are more likely to have voted for the Alliance in 1983 (Heath, Jowell and Curtice, 1985).

All of these variables are used to describe the context within which people are socialised and develop their political ideologies (Scarbrough, 1984). People in different positions within society are differentially socialised, and vote accordingly. Part of the context may be spatial, and it is accepted in two of the books (the exception is Sarlvik and Crewe; see Berrington's 1984 review) that the link between particular attributes and voting may vary from place to place, as a function of the milieux. However, explanations for this variation (as discussed in the next section) are generally weak.

Whereas the first component of the model refers to the long-term processes by which people become linked to particular parties, the second focuses on the voting decision at particular elections: it concerns the process of *political evaluation*. In this case, the focus is on the voting decision at a particular election and the factors which influenced the members of the electorate. There is much debate over this topic — what it is that people are evaluating and how they do it (see Whiteley, 1983, 1986); indeed, one recent study implies that the process is far from rational, since 'How a person votes is a poor guide to what a person thinks about most issues today' (Rose and McAllister, 1986, p. 147). To some analysts, understanding these processes of political evaluation is becoming increasingly important because of the growing dealignment of the electorate (see the essays in Crewe and Denver, 1985; for a brief overview, see Johnston, 1987a); an increasing proportion of the electorate feels no close and enduring ties to a particular party and votes pragmatically at each election, producing an increasing volume of gross inter-party shifts between elections.

For the authors of the British Election Studies, the study of the political evaluation component is almost entirely divorced from that of political socialisation. Thus, for example, in the second edition of *Political change in Britain*, Butler and Stokes (1974) focused on which issues were perceived as important by the electorate, the structure of opinion on those issues, the images of the parties, the evaluation of party leaders, and views about the economy, without closely linking this part of the book to what had been presented earlier with regard to political socialisation. Attitudes and evaluations were studied as topics in themselves, and members of the electorate were simply categorised according to those characteristics and not also according to the contexts within which they were politically socialised and their long-term alignments defined. The same is true of the latest book in the series (Heath, Jowell and Curtice, 1985), although its conclusions do make some attempt to link the two components through the mobilisation activities of the political parties.

Formalising the model

The implicit model derived from the stucture of the British Election Study books and various writings about dealignment can be formalised as a Bayesian function as follows:

$$P_I = (S_I)w_1 + (C_I)w_2 \tag{1.1}$$

where
P_I is the probability of voting for party I;
S_I is attitude towards party I according to the individual's political socialisation;
C_I is attitude towards party I according to the individual's evaluation of it on contemporary issues; and
w_1 and w_2 are weights indicating the relative importance of political socialisation and evaluation in the individual's voting decision.

In this function, S_I, C_I, w_1 and w_2 are measured on scales ranging from 0.0 to 1.0, so that

$$0.0 \le S_I \le 1.0 \tag{1.2}$$
$$0.0 \le C_I \le 1.0 \tag{1.3}$$
$$0.0 \le w_1 \le 1.0 \tag{1.4}$$
$$0.0 \le w_2 \le 1.0 \tag{1.5}$$

and

$$w_1 + w_2 = 1.0 \tag{1.6}$$

Thus, for example, one may have an individual whose political socialisation produces a value of 0.9 for S_I and whose evaluation of party I gives 0.7 for C_I. If that person's decision were largely determined by long-term political alignments, so that $w_1 = 0.8$ and $w_2 = 0.2$, then

$$P_I = (0.9)0.8 + (0.7)0.2 = 0.86,$$

which is a probability of 0.86 of voting for party I. If, on the other hand, most weight were given to evaluation of I on the current issues, so that $w_1 = 0.3$ and $w_2 = 0.7$, then

$$P_I = (0.9)0.3 + (0.7)0.7 = 0.76,$$

which is a much smaller probability of voting for I.

The argument of those favouring the dealignment thesis is that, over time: (1) long-term ties to any particular party have become

7

weaker for a variety of reasons relating to the increased complexity and variability of the political socialisation process, so that all values of S are coming closer to 1/N, where N is the number of political parties available; (2) evaluation of the parties is more pragmatic, so that for any individual the value of C for any party is more variable from election to election; and (3) over time, evaluation is a more important component of the decision-making process than is socialisation, so that the size of w_2 is increasing while that of w_1 is necessarily falling. Unfortunately, putting such a model into operation is not an easy task. Furthermore, it is necessary to disaggregate the model further, in order to make it realistic.

Thus far, S_I and C_I have been presented as single variables. They are, of course, vectors of variables comprising a set of elements that make up the component being analysed. Moreover, the elements of those vectors are not independent; they interact in many ways. Thus, for example, S_I can be subdivided into several elements, including the following:

$$S_I = \left[O_I, H_I, U_I, E_I\right] \tag{1.7}$$

where
O_I is the individual's attitude to party I according to occupation;
H_I is the attitude according to housing tenure;
U_I is the attitude according to union membership; and
E_I is the attitude according to educational status.
Each of these elements will be weighted, so that w_1 has to be subdivided according to

$$w_1 = \left[w_O, w_H, w_U, w_E\right] \tag{1.8}$$

where the individual weights indicate the importance of each aspect of the individual's socialisation to the derivation of the political alignment — i.e. attitude to I. In addition, the elements may of course be interacting: a person who is both a trade unionist and a council-house tenant may be much more pro-Labour in general attitudes than one would expect from summing the probabilities of being pro-Labour for the two elements separately.

Exactly the same disaggregation of the right-hand term of the Bayesian equation (1.1) is possible, since there are many different elements of the evaluation of a party (its policy on various issues; the quality of its leader(s), the nature of its local candidate, its record (if any) in government, and so on) — and, of course, different people will weight these elements differently; some may evaluate a party on the quality of its leader alone, some on its policy on a single issue, and so on.

Finally, there are interactions between the two elements of the equation, since socialisation and evaluation are not independent (at any one election socialisation is more likely to influence evaluation than vice versa, of course, but if socialisation is a continuous process, then over the longer term it is affected by evaluation). They interact, too, because of the role of the political parties in seeking to mould both processes. At national and local scales, these organisations are active in mobilising the electorate, both seeking long-term commitment (for example, through party membership) and promoting their policies; success at one may (and almost certainly should) aid success at the other.

Structuring such a model so that it can be empirically calibrated, with all the probabilities and weights estimated, is a challenging and difficult task (to which Whiteley, 1983, 1986, has recently responded). Such is not the goal here, however, so the model has been presented only as a general organising framework, rather than as a protocol for analysis. The focus for the rest of the book is on one element of one component only — the role of locally provided information as an influence on the evaluation process.

COMPOSITIONAL AND CONTEXTUAL SOCIAL THEORY

The basic thesis of the larger research programme from which this book has emerged is that the study of voting behaviour in Britain must take much greater cognisance of the *geography* of inputs. (The other major study from the research programme so far is Johnston, 1985a, while the reviews in Johnston, 1986a, 1986b, provide a base for the analytical work that is being undertaken. The programme is set in the context of a wider argument regarding the philosophy and content of human geography: see Johnston, 1984, 1985f, 1986d.) This involves recognition of two types of social theory, and the need to integrate the two.

The first of these types, and by far the most common, is *compositional theory* (see Thrift, 1983), the basic premise of which is that people's behaviour can be derived from knowledge of certain of their individual characteristics. This has clearly been the case in much British psephology, for party alignment has been closely associated in much writing with class position, particularly as reflected in occupational class. Thus, for example, Bogdanor (1983, p. 53), writing about the decades immediately after the Second World War, stated that

> It was class voting which gave to British politics its electoral stability. Electoral behaviour came to display a considerable degree of geographical homogeneity since an elector in Cornwall would tend to vote the same way as an elector from a similar class in Glasgow regardless of national or locational differences.

The implication is clearly of a national political culture: all people occupying a particular niche in society are socialised in the same way. Human agency is involved in making the decision regarding political alignment, of course, but the decision-making is structured similarly in all parts of the country. This is an argument taken further by Dunleavy and Husbands (1985), as follows: (1) people occupy social locations, reflecting their role in the production activities of society and their position with regard to the consumption of goods and services; (2) their political interpretations of those locations reflect the messages that they receive via the mass media; (3) those interpretations are used to evaluate competing political parties in which they usually act collectively:

> most people most of the time act instrumentally to further the interests of their social location. They do not undertake an analysis of their individual household situation but rather act to promote the collective interests of their social location, as these have been defined in their society. (p. 20)

Thus social location is not just a position within society; it is a focus of political mobilisation.

The polar opposite of this compositional type of theory is *contextual theory*, which characterises people not by what they are, but rather by where they are. The crudest expression of this within geography was the theory of environmental determinism, which held that the environment controls the course of human action. Human agency is thus denigrated, and people are seen as simple products of their natural environments. Such simplistic reasoning has long been discredited, of course, but it still influences much geographical work, albeit implicitly (Johnston, 1986d). More importantly, a substantial volume of work is built on the foundation of the assumption that who you live among is a major determinant of your behaviour (as exemplified in Miller's statement quoted above, p. 3).

Neither the compositional nor the contextual approach is satisfactory in itself: a fusion of the two is necessary. Compositional theory is unsatisfactory because it ignores the contextual (i.e. locational) framework within which all behaviour occurs. It could be argued that statements like those of Bogdanor, and Dunleavy and Husbands, quoted above, do not fall into this trap, because they assume a national context. True, but within that national context they assume uniformity, an assumption that is difficult to sustain (Taylor, 1985a; Johnston and Taylor, 1986). Dunleavy and Husbands, for example, refer to voters promoting 'the collective interests of their social location, as these have been defined in their society'. But what is 'their society'?; for manual workers who are council tenants, is it all manual workers who are council tenants in the UK? Taylor's arguments about the scale of experience, and a wide range of other writings on people's lifeworlds, suggest not, but rather, that most people are restricted to relatively small spatial ranges (what Giddens (1984) terms locales). It is in these contexts that they learn what membership of particular social locations means, and there is no guarantee that the same set of meanings will develop everywhere (see Johnston, 1986c). As Przeworski (1985, p. 99) puts it:

> Class, ethnicity, religion, race or nation do not happen spontaneously, of themselves, as a reflection of objective conditions in the psyches of individuals. Collective identity, group solidarity, and political commitments are continually forged — shaped, destroyed, and moulded anew — as a result of conflicts in the course of which political parties, schools, unions, churches, newspapers, armies, and corporations strive to impose upon the masses a particular view of society.

Some of those agents — particularly those directly linked to the state and its role in the development and promotion of a national ideology — may operate at the national level; but many are implicated at the local level, in the development of a mosaic of local cultures within the broader context of a national culture.

A theoretical position which brings together the compositional and the contextual, with much greater stress on the contextual than provided by most social theorists, is *structuration theory*, very largely the product of one sociologist, Anthony Giddens (see Giddens, 1984). In this approach, Giddens distinguishes between social systems, which are *situated practices* — 'patterns of relationships between actors or collectivities reproduced across time and space' (Giddens, 1981, p. 26) — and *structures*, which are 'the medium and the outcome of the practices which constitute social systems' (p. 27). Thus, structures guide action, and are reproduced (perhaps in slightly altered form) by action. Because of the spatio-temporal constraints to much of life, those structures are spatially (and temporally) restricted; they form locales, settings for interaction that are the milieux within which people learn and act.

Structuration theory is now the subject of a large literature, which will not be explored here. Its importance to the present work is the theoretical underpinning that it provides for the argument that the meaning of one's social location is very largely appreciated in one's contextual location; one learns about what one is in the context of where one is. This is clearly relevant to the long-term processes of political socialisation and to the findings of many analysts that knowledge of people's social locations is insufficient to account for their voting behaviour; people in similar social locations interpret them differently in some spatial locations from those in others. (For a recent debate on this issue with reference to voting in Britain, see Johnston, 1987b, and McAllister, 1987.) It is also relevant, it is argued, to the processes of political evaluation, since it is in individual places, as well as nationally, that parties and their candidates seek to influence voters. It is the nature of that influence, and its impact on how people vote, which constitutes the research contribution of the present book.

ELECTORAL CAMPAIGNING

Political parties are institutions created in order to advance both general and specific goals, relating to the actions of government and the individuals who control governments. They are sustained by the support which they receive, both from the electorate and from various sponsors (most of whom are also members of the electorate). Some parties, because they are firmly linked to particular goals, remain in existence because they have the support of a defined portion of the electorate (i.e. people in particular social or contextual locations, or both); but most are pragmatic because their major goal is to attain elected power (see Taylor, 1985a, 1986; Johnston, 1987a), and they devise strategies in order to promote those goals. Such electoral strategies can be divided into two major components:

the long-term processes of political socialisation via mobilisation, and the short-term activities of political persuasion. The former involves continuous activity; the latter — which is the focus here — is especially important in the period prior to an election, especially the final weeks when many voters make their decisions. This process of persuasion involves electoral campaigning.

The goal of an electoral campaign is to convince as many people as possible to vote for the relevant party/candidate. It involves a strategy of both designing a programme with which to sell the party/ candidate (in general advertising terms, 'creating a product'), packaging that programme (creating an image for the product), and then selling it (marketing the product). In most British election campaigns, this involves, at the national level, the production of a party manifesto, the selection of leaders, and a programme of events designed to present manifesto and leaders to the electorate at large; it involves both retrospective (what has been achieved) and prospective (what will be achieved) components, and contrasts its programme and leaders with those of the opposition. At the local level, it involves the selection of a candidate and of local issues which mesh with the general policies of the party, and putting these together as part of a general/local package designed to promote the party nationally through its local representative.

The national and local campaigns differ in terms of their operation: nationally, contact with the electorate can only be made through the mass media; locally it is feasible to make more personalised contact, in one of four ways: (1) by either the candidate or a party member visiting the individual electors, in their homes, at their workplaces, or in some other locations (e.g. shopping centres); (2) by holding meetings, at which electors can hear and question the candidate; (3) by advertising, either visually (posters) or orally (using loudspeakers); and (4) by delivering publicity material (perhaps personalised) to the voter's home (see Rose, 1976, Chapter 3). The rationale for all of this activity is clearly to inform and to convince; it may be bolstered by canvassing activity, designed to identify those likely to vote for the party and then to ensure that they cast that vote (see Bochel and Denver, 1971, 1972).

Most of this activity requires money. Creation of a manifesto involves research by permanent staff and by contracted workers (e.g. survey researchers); its promotion requires printing and publicity, including, if possible, buying and using media space and time, and paying the expenses of the leader and other members in a variety of campaigns. A permanent administrative presence is necessary for maintaining all of that activity, and for responding to issues as they arise. For the local campaigns, preparing the publicity material involves costs, as will hiring the facilities for meetings (if permanent facilities are not already owned). Much of the work may be done by voluntary activists, but a permanent organisational presence (e.g. an agent) is desirable to co-ordinate all of this activity and provide continuity. Both maintaining contact with and servicing these local organisations places further demands on the central administration.

Political parties in mass democracies are not viable if they lack money, since they have no way of contacting the electorate, let alone developing policies to present to the electorate, unless they can buy both labour and materials. (Some labour will be available free, but insufficient for the purposes outlined here.) Money can be obtained from three sources: (1) from party members; (2) from sponsors, who are not members; and (3) from state subsidies. The volume of the first depends upon the number of members, their affluence, and their willingness to give; that of the second depends upon whether wealthy sponsors (individual or institutional) are prepared to donate, either out of altruism or, more likely, from an expectation of benefits, should the party/candidate sponsored be successful; and that of the third depends upon state attitudes and, if these are favourable, the criteria for eligibility (these apply to subsidies in kind — e.g. free broadcasting time — as well as money payments).

It is generally assumed, in line with research findings, that advertising helps to sell a product, either increasing demand for it or increasing the market share of a particular provider. If it is assumed that political campaigning can be equated with advertising (as in the writings of Palda and Chapman reviewed below, p. 26) because it involves the provision of information designed to increase the vote share of the party/candidate concerned, then to the extent that parties and candidates differ in the amount of money that is spent on this activity, so the result of an election will, in part, be a function of the volume of spending.

Two aspects of the role of campaign funds in elections have concerned many people: the inequalities between parties (especially between established and new parties) in the amounts which they are able to raise and spend on the campaign, thus having a differential impact on the electorate; and the ability of wealthy sponsors (individual or corporate) to invest massively in a campaign, thereby dominating the flow of information and substantially influencing the election result. (A third aspect — bribery — is possible, but is largely outlawed; pork-barrel politics — Johnston, 1980a — whereby public money is directed to particular constituencies in the search for electoral gain, is equated with bribery by some.) To counter these phenomena, many countries have introduced limits on the amount that can be spent and on the size of individual donations, requiring disclosure of all sums received and spent, and backing up the controls with legal enforcement. (For a review of such legislation, see Paltiel, 1981.) Limits on the size of contributions are designed to ensure that small numbers of individuals/organisations do not control the content of a campaign and the subsequent political response (i.e. the geography of outputs); spending limits ensure that those less able to raise campaign funds are not too disadvantaged. Both can, of course, be circumvented in a variety of ways, so that some countries have introduced either public subsidy or a total financing of campaigns (Paltiel provides a listing of these).

The United Kingdom has not been immune from these concerns, and has reacted with legislation designed to limit the amounts spent and to provide subsidies for the parties. The concern remains, however, and in 1975 a committee was established by the government to inquire into the desirability of state aid for political parties (Houghton Report, 1976). A majority of its members recommended that such aid be provided, but this was never acted on by the (Labour) government of the time.

One aspect of the political-party financing issue which has not been tackled — either by the politicians concerned or by political scientists — is its impact on the election results. For money spent on the national campaign this is very difficult to assess, since it is hard to estimate the counterfactual situation; that is, what would the result have been if the pattern of spending had been different. For constituency campaigns an assessment is possible, however, not because a counterfactual can be established, but because it can be argued that if the amount of spending varies across the constituencies, then whether this variation is related to the pattern of voting, *ceteris paribus*, can be tested statistically. The evaluation is statistical only, of course, so cause has to be inferred in the context of a model indicating why such an inference is valid. That model has been provided here with reference to the processes of political evaluation (p. 6). Parties spend on constituency campaigns in order to promote their policies and candidates among the electorate. Assuming that people do not vary, on average, between constituencies in their degree of receptivity to such information, then it is reasonable to assume that the more information is provided, the more voters will respond to it. This is the basic hypothesis underlying the analyses reported here.

POLITICAL FINANCE IN BRITAIN

The regulation of political finance in Britain commenced in the nineteenth century as a response to a variety of electoral abuses, notably bribery and 'treating' (the wooing of voters by the provision of gifts and favours rather than direct sums of money). Thus, the Corrupt Practices Act of 1854 was introduced to outlaw bribery of voters, with bribery being defined so as to include both treating and the offer of either public office or employment (see Seymour, 1950, p. 229). There were still many corrupt practices decades later, however (as detailed by both Seymour, 1950, and O'Leary, 1962), and this led to the introduction of the Corrupt and Illegal Practices Act of 1883, characterised by Seymour (p. 455) as 'a landmark in the development of democracy in England . . . classed with the legislation of the two succeeding years, which extended the franchise and reformed the distribution of electoral power'.

The main features of the 1883 Act, as far as the present study is concerned, concern its imposition of spending maxima in each constituency, the rules that were provided to govern the spending of moneys during the campaign, and the introduction of clear penalties (a fine of £100, exclusion from voting rights and public and judicial offices for five years, and seven years' exclusion from their constituency for those candidates convicted personally for illegal practices). There were eight categories of constituency, each with its own spending maximum (see O'Leary, 1962, p. 175) as follows:

English, Scottish and Welsh boroughs:
    with less than 2000 voters — £350
    with 2000 or more voters — £380 plus £30 for each extra 1000
English, Scottish and Welsh counties
    with less than 2000 voters — £650
    with 2000 or more voters — £710 plus £60 for each extra 1000
Irish boroughs
    with less than 500 voters   — £200
    with 500–1000 voters   — £250
    with 1000 or more voters — £275 plus £30 for each extra 1000
Irish counties
    with less that 2000 voters — £500
    with 2000 or more voters — £540 plus £40 for each extra 1000

A clear differential was established between the borough and the county constituencies, on the grounds that contacting voters was a much more expensive task in rural than in urban areas. The rules governing expenditure requires that all contracts and payments had to be authorised by the candidate's agent, set a further maximum of £100 for the candidate's personal expenses, and required a detailed return within 35 days of polling.

The immediate consequence of this Act was a reduction in the amount spent, according to the candidates' returns, from an average of 18s 9d (94p) in 1880 to 4s 5d (22p) in 1885. Virtually the maximum allowed (98 per cent) was spent in England and Wales in 1885, but ten years later only 80 per cent was expended. Thus, not only were bribery and (as far as was possible) treating very largely eliminated but, according to O'Leary (1962, p. 206), 'Even if other methods were available for spending money in politics, the actual cost of electioneering was reduced by three-quarters, which was a great advantage to the poorer candidates.' (Whether this advantage was reflected in the voting returns is not stated.) Indeed, it was the size of the maximum and its effect on the poorer candidates and parties which were central to the debate over election expenses in the following decades.

During the twentieth century, the extension of the franchise meant that the number of electors to be contacted by parties and candidates grew very rapidly. Under the Representation of the People Act, 1918, the maxima were changed to 5d (2.08p) per elector in borough constituencies and 7d (2.92p) in county constituencies, which, as Butler (1963) shows, meant a reduction of the maxima to less than half of the 1883 sum. Subsidies were also introduced, however, notably one free postal delivery to each elector and the free use of school halls for meetings; and a payment to an agent was allowed. The limits were further reduced in 1928 under the Equal Franchise Bill — with the county figure changed to 6d (2.5p) per elector. Initially the borough figure was to be reduced by 1d also, but this was defeated. Labour MPs were in favour of the change but, as Butler (1963, p. 55) describes it

The Conservatives were divided on the issue. On one side they argued that they needed more money than Labour to put their case adequately before the electorate, as they lacked the free trade-union publicity and clerical assistance which Labour had . . . On the other side, they argued that if there was a high maxima, pressure would always be brought upon them to spend to that limit: high election expenses often deterred their best men from standing.

Further attempts were made by Labour MPs to reduce the maxima as part of the bargaining over electoral reform in 1930, but the bill fell with Ramsay Macdonald's Labour government. (See Butler, 1963, pp. 58–83, for a detailed discussion.)

The question of the maxima was raised again in 1944 following a Speaker's Conference, which proposed that the maxima should be £450 plus 1d (0.42p) per elector in borough, and 1.5d (0.63p) in county constituencies. This unanimous recommendation would have cut the maximum spending allowed per voter by nearly one-half, and led Butler (p. 95) to state: 'It is curious to speculate upon what had convinced the Conservatives that beyond a certain point the expenditures of money did not help to win elections'. Certainly it was not research that was later published. He continues thus:

> It appears . . . that pressure for lower expenses actually came from the Conservatives. During the war the Labour party and Trade Union funds had grown very considerably and much more money was available for electioneering. The Conservatives, on the other hand, being accustomed to look to their candidates financing their own campaigns, felt much poorer. High taxation would make it impossible for many private individuals to pay their expenses on the pre-war scale. A movement to take all expenses off the candidate's shoulders was under way, but central party funds were also limited at that time. Everything led the Conservatives to wish for a considerable reduction, while the Labour party, feeling more affluent than formerly, and conscious of the need for campaign publicity in order to counter a predominantly Conservative press, were perhaps not so eager for change as they had been (pp. 95–6).

No changes were introduced until the Representation of the People Act of 1949, however, which once again reduced the maxima, to approximately the levels proposed in 1944 (i.e. £450 plus 1.5d per elector in borough constituencies and 2d in counties). There was little debate on the issue. This was the last change for 20 years, although in the interim period national party expenditure increased substantially, together with much related expenditure by businesses and other interested groups.

It is the 1949 Act which remains the basic document on what can be spent by candidates during an election campaign. As Pinto-Duschinsky (1981a) points out, the Act refers only to expenditure incurred in promoting an individual's candidacy. Before the officially defined campaign period (i.e. between the declaration of the election and polling day), expenditure can be justified as promoting the local party association rather than the candidate: after then it cannot be so justified. The Fifth Schedule to the Act indicates the categories of expenditure that are to be reported: candidate's personal expenses; payments to agents and sub-agents; payments to clerks and messengers; payments for printing, advertising and stationery; payments to speakers for expenses; payments for rooms for meetings and for committee rooms; payments for postage and telegrams; and payments for miscellaneous matters (with reasons for the payments). All bills and receipts are to be submitted with the return.

Pinto-Duschinsky (1981a, p. 249) argues that despite these constraints the law is not watertight:

An experienced election agent can normally find ways of stretching permitted expenditures. Printers are persuaded to give low quotations for electoral literature with the promise of further business after the campaign. Paper is purchased before the start of the campaign on behalf of the local party association and then sold to the candidate as second hand stock. Election agents' fees are frequently artificially low. These devices can, where required, provide an extra 20 per cent of expenditure. It is not possible, however, to abuse the limits on a large scale without risking the draconian penalty of having his election declared void.

Thus, to the extent that agents differ in their ability to use these devices, so the returns of expenditure vary, by unknown percentages, from the 'real' expenditure. (See also Gordon and Whiteley, 1980.) For the analyst, however, the published returns have to be taken as the best available.

The period studied here is that since 1950, entirely within the confines of the Representation of the People Act (1949), as amended. Earlier elections were excluded for a variety of reasons, including the existence of two-member seats until 1950 and the lack of candidates from even the main parties in many constituencies before 1939 (for example, because of the Lib-Lab pact). Since 1951, the first election analysed (see below, p. 61), the majority of the spending has been on publicity materials (Table 1.1), with the percentage of the total spent on that item increasing over the period with the declining importance of public meetings and the smaller number of agents and clerks employed. To all intents and purposes,

Table 1.1: The percentage distribution of campaign expenditures

	1951	1959	1964	1966	1970	1974	1979	1983
Agents	8	8	7	7	6	5	5	5
Clerks etc.	6	6	5	5	4	2	2	2
Printing/Stationery	67	69	72	72	75	80	80	79
Public Meetings	6	3	3	3	2	2	2	2
Committee Rooms	5	5	5	5	4	3	3	3
Miscellaneous	9	10	9	9	9	7	8	9
Average personal expenditure per candidate (£)	37	40	41	39	41	42	51	65

therefore, the analysis of spending is the analysis of advertising, the purpose of which is to inform and persuade.

Much of the debate over the spending maxima has concerned the differential abilities of parties to raise funds. As indicated above, immediately prior to the period to be studied here the Labour party, with its links to the trade-union movement, was considered to be in a better position to raise and spend funds than the Conservative party, whose candidates largely financed their own campaigns. The local Liberal parties were generally poor. Since 1951, as the analyses to be reported later indicate, the Conservative party has been the best able to finance local campaigns overall. This is largely because of the results of local efforts for, as Pinto-Duschinsky (1981a) points out, only 2 per cent of the party's expenditure in the 1970s went in grants to constituencies, and this accounted for only about 3 per cent of total expenditure there: branch subscriptions and money raised at a variety of social and other events contributed the main source, and most of the grants from the central office were targeted on the Labour-held marginal seats (Pinto-Duschinsky, 1981a, p. 142).

Local Conservative parties have clearly become very efficient and effective at raising campaign funds in recent decades but (Pinto-Duschinsky, 1981a, p. 155) 'The main contrast between Labour and Conservative finances since the Second World War has been Labour's relative failure in raising small-scale funds at the constituency level', despite relative prosperity for the central organisation. As a consequence, the Labour party has felt much more obliged to provide grants to weak constituency parties, amounting to 16 per cent of the central budget in 1979 — although on average a constituency party only received about 2.5 per cent of its income from this source; gambling schemes provided most income. The Liberal party has always relied heavily on local efforts to raise funds (although a Direct Aid Committee was established in the 1970s to provide grants to 'constituencies already held by Liberal MPs or those considered most promising' — Pinto-Duschinsky, 1981a, p. 186). Nevertheless, a substantial proportion of the party's relatively small central budget has been disbursed in grants to constituency parties (43 per cent in 1974). However, in making the grants, the party has been caught between: (1) wanting to bolster its vote in the few seats it held together with those it could win; (2) wanting to ensure that as few candidates as possible lost their deposits; and (3) wanting to win votes throughout the country to promote the case for electoral reform. Thus, central grants have been but a small part of total constituency expenditure. The Liberal party's 1983 Alliance partner, the Social Democratic Party (SDP), operated differently, however. It was much more centralised, with members' subscriptions going directly to the central organisation. A substantial proportion of total expenditure (about 50 per cent) spent on the 1983 campaign went on grants to constituency parties, however (Pinto-Duschinsky, 1985); all candidates were assured of receiving at least £2000, and incumbent MPs received as much as they needed, up to the maximum.

The general situation with regard to expenditure on local campaigns in Britain, therefore, is as follows:

(1) The great majority of it is spent on advertising, as a means of influencing the process of political evaluation; and

(2) In real terms over the period studied here, the amount spent has declined substantially. Pinto-Duschinsky (1981b) shows that in constant (April 1979) £s, the amount spent per candidate fell over the period 1945–70 by 66 per cent for Conservative, 61 per cent for Labour, and 77 per cent for Liberal. (This decline in expenditure is at least partly countered by the subsidies-in-kind, which at the local level are the free postage and the free use of halls. These have been available throughout the period, however, and there is no evidence that they have increased in relative terms. The main subsidy-in-kind is the free broadcasting time.) From this situation one could infer that the parties are placing less importance on local advertising, since they are not seeking to increase the maxima and are not, in most cases, spending as much as they could. The implication is that they believe that local advertising is having a declining impact on the process of political evaluation — indeed, they may believe that it has none at all, and simply undertake it as a matter of routine. However, that implication has never been put to the empirical test, and such is the aim of this study.

IN SUMMARY

Voting is presented here, in line with much recent British psephological writing, as a two-component decision-making process. The first component involves the long-term political socialisation of people, which stimulates the creation of an individual ideology linking the voters to particular parties. The second component involves the short-term processes of political evaluation whereby voters, in the context of their long-term socialisation, evaluate the political issues, events and personalities of the recent past and formulate views about the best policies and individuals for the likely issues of the immediate future. Some will have their decision on how to vote determined largely by the former process, others by the latter, and others still by a more balanced mixture of the two.

Both political socialisation and political evaluation are processes that occur in places; people are influenced both in their long-term development and in their immediate decision-making by elements of their milieux (some more than others, of course). Thus, in order to appreciate fully the results of an election, it is necessary to appreciate the milieux in which people both have learned about, and go on learning about, politics.

One aspect of those milieux is studied here — the politically relevant information circulated by parties during election campaigns in order to try and influence voters and win their support. Such information circulation is akin to advertising, and should have similar results: the more that a competitor spends relative to others, the greater should be the resulting market share. Whether that is a valid statement is the subject of the research reported here, conducted in the context of British electoral law and with particular reference to general elections in England since 1950.

2

Spending and Votes: A Review

Although party finance has been a regular if infrequent topic of political debate in Britain, most of the discussion has focused on differences in the money available to political parties: little has been said on how it has been used, and to what effect. Certainly there have been very few analyses of the not insubstantial sums of money spent by constituency parties during general election campaigns, and yet that money is expended for the single reason of seeking an increased share of the votes cast. Thus, if parties are differentially able to raise money for their campaign activities, and the amounts spent do have an influence on the results, then the availability of funds for campaigning is an important influence on the outcome of an election. To the extent that spending and votes are related, democracy is subject to the 'power of the purse'.

Most analyses of the link between spending and vote-winning — as distinct from general writings based on anecdotes and unsubstantiated generalisations — has been undertaken in the United States. To inform the present analyses of Britain, that literature is briefly reviewed here. The small amount of work on spending and votes in Britain is then summarised, providing a basis for the development in Chapter 3 of a model to be evaluated in the remainder of the book.

CAMPAIGN SPENDING AND VOTES IN NORTH AMERICA

No attempt will be made here to provide an exhaustive coverage of the literature on the relationships between spending activities in North America. Rather, a selective review will be used to show the developments in modelling which have been achieved there.

Money and votes: a production function

The first study of relevance here was by Welch (1974), who investigated the 1972 elections to both houses of the US Congress. He noted that although much work had been done on the financing of political campaigns, little had been reported on the impact of that money — what he termed 'the transformation of economic power into political power' (p. 83). Only three previous empirical studies were identified, and they were categorised as both lacking in theory and statistically unsophisticated. He suggested that campaign contributions will be demanded by candidates because they can be translated into an expected vote percentage: the more that a candidate spends, *ceteris paribus*, the greater his or her share of the vote, whereas the more than the opponent spends, the less the candidate's share. Thus, very simply, candidates are competing for a share of the electorate, and the greater the relative spending of one, the greater the share that person should win. No campaign is held in a vacuum, however, especially one for the US House or Senate. Most of the electorate has voted before and is associated with a particular party; voters will have developed voting habits, the sum of which comprises the 'normal vote' (Converse, 1966) for that area. The goal of the campaign spending is to produce deviations from that normal voting behaviour. Furthermore, Welch notes that incumbents have immense advantages in the American system, through media coverage, government subsidies available (free use of the postal system, for example), and ability to raise money. Thus, Welch formulated a multiple regression function, as follows:

$$RV = f(RE, DE, RI, DI, NV) \qquad (2.1)$$

where
RV is the Republican percentage of the vote;
RE, DE are the Republican and Democrat candidates' expenditures, respectively;
RI, DI are dummy variables, set at 1 if the relevant candidate (Republican and Democrat respectively) is an incumbent; and
NV is the normal vote for the Republican party.
For the Senate election in 1972, the equation accounted for 66 per cent of the variation, with only the two expenditure variables significantly related to the election outcome; the more that the Republican candidates spent, the more votes they got, whereas the more that the

Democrat candidates spent, the less votes the Republicans got. In the House elections, all five independent variables were significantly related to the dependent (with the expected signs), and they accounted for 81 per cent of the variation. In both cases, taking the logarithms of all the non-dummy variables slightly increased the goodness-of-fit; the interpretations were the same, except that in the case of the Senate, a Democrat incumbent had a significant negative impact on the Republican performance.

Welch (1976) extended his work to research on State legislature elections in California and Oregon, including primary elections. In addition to the hypothesis tested in the original paper, he also proposed: (1) that spending would have a greater impact in primary elections, when party label was irrelevant, since the contest was intra-party; and (2) that money would have a greater impact in large than in small districts because of economies of scale in its use (especially in the purchase of broadcasting time). The results relating spending, incumbency and the normal vote to the election result were generally as expected and the equation fits were good. The additional hypothesis regarding primary elections was verified, but that relating to district size was not.

In this second paper, Welch argued strongly for the use of the equations in their logarithmic form, because this is consistent with the Law of Diminishing Returns. As he concludes, 'the marginal product of money decreases rapidly with increased expenditure . . . the more money spent by a candidate, the less the vote percentage "bought" by an added dollar' (p. 353), with the implication that large subsidies to campaigns should be provided from public funds; additional money raised by the candidates (from interest groups hoping to benefit from the election of the person supported) would then have little influence, and there would be no incentive for raising campaign funds.

Similar models have been used by Patterson and Caldeira in a range of studies. For the 1978 gubernatorial elections, for example, they showed a close relationship between the percentage of all spending associated with one candidate and that person's share of the votes cast. Indeed, it seemed that the impact of spending was very substantial: '. . . other things being equal a 10 percent increase in the proportion of campaign spending by a candidate yields a 5 percent increase in the percentage of the general election vote for governor' (Patterson, 1982, p. 463). However, a candidate whose party was already strong in a State tended to spend more, as did incumbents, so that once these influences had been taken into consideration, the impact of spending may have been reduced. This was not so, however; *ceteris paribus*, the more that the Democrat candidates spent, the more votes they received (though there was clear evidence of diminishing returns for additional expenditure), and the more that the Republican candidates spent the less votes the Democrats obtained (again, with evidence of diminishing returns). Furthermore, spending not only influenced the share of the vote, it also influenced the turnout —

> Variations in spending represent variations in campaigning effort — candidates and parties spend money to advertise, canvass, hold rallies, promote, conduct polls, and otherwise endeavor to get at the vote. Campaign spending pays off very impressively in increased turnout at the polls (Patterson and Caldeira, 1983, p. 685)

— though, again, with diminishing returns. Similarly, Caldeira and Patterson (1982a) reported that spending had a positive impact on turnout rates in California and Iowa State elections up to a certain level, beyond which diminishing returns set in; beyond a further threshold, in some cases extra spending apparently led to a decline in turnout. (For the Iowa Senate, for example, they reported that in the late 1970s, below the first threshold every $1000 spent by the parties apparently boosted turnout by 1.4 percentage points.)

Gross campaign-spending, according to these findings, influences the level of turnout. The proportion of that spending undertaken by a particular party influences its share of the vote. Analyses of California State elections have show that 'money is the best predictor of the vote, party strength is next in order of importance and incumbency adds the least in explanatory power' (Owens and Olson, 1977, p. 511), leading to the conclusion that 'money does win elections and statutory limits must be placed on spending' (ibid). Owens and Olsen also noted that incumbents were better able to raise campaign funds (a point taken up below: p. 29), so that incumbency had an important indirect influence on the outcome (see also Tuckel and Tejera, 1983).

Analyses of Canadian elections have produced very similar results (as in the initial studies by Palda, 1973, 1975). Chapman and Palda (1981) have set these in the context of a public-consumption perspective, arguing that the likelihood of a person voting is a function of the probability that each vote will influence the outcome, *ceteris paribus*. (The latter included an allowance for the well-attested finding that better-educated, better-paid individuals are more likely to vote.) For the individual, evaluating that probability means obtaining information about the contest. If such information is provided free, by the contestants, the cost of acquiring information falls and the likelihood of voting increases; making the voting decision is made less 'expensive' in terms of effort by campaign spending. However, analyses of ten provincial elections produced a significant positive relationship between turnout and spending in only three (two in Ontario), together with an unaccountable significant negative relationship in one other (Chapman and Palda, 1983).

Incumbents and challengers

The studies reviewed in the previous section have tested a very straightforward set of models which state, *ceteris paribus*, that: (1) the more that is spent in a campaign, by all candidates, the greater the turnout rate; and (2) the more that each candidate spends, the greater the resulting proportion of the votes won. With regard to the latter model, tests which introduced additional variables showed that incumbency had both direct and indirect impacts on the outcome: incumbents tended to win more votes; they also tended to spend more. The role of incumbency was not very fully explored, however.

A major development on the second model was provided by Jacobson, who looked at the impact of incumbent and challenger spending separately. He has summarised studies covering the decade 1972–82 (Jacobson, 1985, p. 13) as showing that

> In contests involving incumbents, the more a challenger spends, the greater his share of the vote. The more the incumbent spends, on the other hand, the smaller his vote . . . This does not mean that incumbents lose votes by spending money, but rather that they spend more the more strongly they are challenged, and the stronger the challenge the worse for the incumbents.

The reason for this can be found in the rational voting models of Chapman and Palda. Surveys of voters' ability to recall and recognise candidates' names have shown that they are much more likely either to recall (unprompted) or to recognise (prompted) an incumbent's name than a challenger's, especially in contests for the

House (Jacobson, 1983, pp. 88–9). This is simply because of information, which suggests to Jacobson why incumbents generally do well in US elections: 'voters are much more likely to remember their names' (p. 87), because incumbents have so many media and other resources available with which to promote themselves. Much of this is self-generated:

> House incumbents normally do not attract much attention from the news media. This means that, except during campaigns, they produce and disseminate much of the information about themselves that reaches the public. To a large extent, they control their own press: no wonder it is a good press, and no wonder voters tend to think highly of them (p. 95)

Their challengers must counter this dominance of the information channels by disseminating their own materials, through campaign spending. The more that challengers spend, therefore, the less should be the inbuilt advantage of the incumbents.

This hypothesis has been tested in a variety of ways. Jacobson (1983) used survey data, in which the dependent variable was respondent's familiarity with the candidate: this was scaled at 1 if the name was recalled, 0.5 if it was recognised but not recalled, and 0 if it was neither recognised nor recalled. He found that the amount spent by incumbents had no influence on familiarity, but that the amount spent by challengers did (as did the amount spent by both candidates in open seats, where no incumbent was standing). Goldenberg and Traugott (1985) produced broadly similar results, showing how money is translated into recognition. For challengers in the 1978 House elections, those who spent less than $25,000 were recognised by only 27 per cent of respondent electors, whereas those who spent $100,000 or more were recognised by 70 per cent. Incumbents also benefited, but not to the same extent: the respective percentages were 78 and 87. Thus, spending by challengers reduces the 'information benefit' held by incumbents. Where the challenger spent less than $25,000, only 26 per cent recognised the names of both challenger and incumbent, whereas 54 per cent recognised the name of the incumbent alone; where the challenger spent at least $100,000, the percentages were 66 and 17 respectively. Jacobson interpreted his findings as showing that in 1978 the average challenger would have to spend $320,000 to become as familiar with the average voter as the average incumbent; the average incumbent spent $125,000, however, whereas the average challenger spent only \$72,000 (Jacobson, 1984). Hence his conclusion that

> explains why campaign money is crucial to challengers and other nonincumbent House candidates. Without it, they are likely to remain obscure and so to be swamped by the opposition. They also explain why incumbents receive little measurable benefit from campaign expenditures. The campaign adds little to the prominence and affection they have gained prior to the campaign by cultivating the district and using the many perquisites of office (Jacobson, 1983, p. 106).

Since, as the 1978 spending data show, most incumbents raise much more money than challengers, it is not surprising that the spending by the latter has a greater impact on the result. For 1980, Jacobson (1984) showed that for every \$10,000 spent by a challenger, the vote percentage was increased by 0.35; for incumbents, there was no significant relationship between spending and votes. (See also Copeland and Patterson, 1977.)

Raising and spending

The differentiation between incumbents and challengers provides a much better understanding of the impact of spending on votes than do the simple undifferentiated models, as shown by Jacobson's (1978, 1980) detailed analyses. However, a further issue has to be addressed. As Welch (1981, p. 209) expressed it, 'not only does the amount of money spent probably affect the ballots cast, but also the votes that a candidate is expected to receive presumably affects the money contributed to him and hence spent by him'. People contribute to campaigns in order to assist the candidate's search for victory, from which they anticipate some benefit, either in a general sense or, in the case of institutional donors, in terms of specific advantages (see, for example, Welch, 1982). They are more likely to contribute if they think that the money will help — if the candidate has a chance of winning. Thus, how much a candidate spends will influence the electoral outcome, but how much is received will reflect the expected electoral outcome. The best estimate of the latter is probably the previous election outcome, giving the following simplified model:

$$CE_t = f(CV_{t-1}) \qquad\qquad (2.2)$$

29

$$CV_t = f(CE_t) \tag{2.3}$$

where

CE_t is candidate expenditure at election t;

CV_t is candidate share of the vote at election t; and

CV_{t-1} is candidate share of the vote at the previous $(t-1)$ election. (See also Chapman and Palda, 1984; Caldeira and Patterson, 1982b.)

This issue was recognised by Jacobson (1978), who accepted that the results of the models reported above could be based on unreliable regression coefficients. The apparent impact of spending at election t could be no more than a reflection of the result at election $t-1$; candidates who did well at the first election would attract more funds, but they may have done well at the second without those funds. To circumvent this problem he used two-stage least squares estimating procedures (see Todd, 1980). The results not only confirmed, but strengthened the original findings: 'Spending by challengers has a much more substantial effect on the outcome of the election even with simultaneity bias purged from the equation' (Jacobson, 1978, p. 275) — so that the relationship of spending to votes at election t was not simply a reflection of the relationship between votes at election $t-1$ and spending at t; spending had a 'real' impact on the election result. Thus, Jacobson (1980, p. 162) summarises his major study as showing that 'candidates are given money according to how well they are expected to do, but campaign expenditures have an independent effect on how well they actually do, because without them, the expectation would not be realised'.

That effect of expenditure on outcomes referred only to challengers' spending, however. As Jacobson (1985, p. 41) expressed it,

No matter how the data are analyzed . . . one finding remains undisturbed: incumbents gain nothing in the way of votes by spending money in campaigns . . . If incumbents really gain nothing by spending money in campaigns, ironies abound. Incumbents spend defensively and reactively, but pointlessly . . . The unpleasant work of fundraising which most members of Congress complain of so passionately is not even necessary.

Can this really be so? Jacobson explores two possibilities.
(1) The returns to incumbent spending are tiny but positive, and are not decipherable in the 'noise in the data'. Most of the small number of incumbents who lose do so in close contests, in which small shifts

in the result could be crucial, and could be influenced by spending. Such a hypothesis is not readily testable, however.

(2) Incumbents invariably spend lavishly when they face a serious challenge, so that the effect of low spending by incumbents is rarely investigated. The small amount of available evidence does suggest that incumbents who fail to match a well-funded challenge are more likely to lose than are those who equal the challenger's spending. There is, then, an apparent threshold effect.

Jacobson favours the second option. For the non-incumbent, there is a threshold that must be crossed in order to provide information for a creditable challenge. The incumbent has no need to spend a matching sum to maintain visibility, but may have to if the issues require a vigorous campaign against the positions taken up by the challenger (Jacobson suggests this is most likely at times of low government popularity). Thus, for the incumbent the saliency of spending reflects the content of the campaign, not its volume. Aggregate data cannot capture such qualitative differences so that analyses rarely show a positive relationship between incumbents' spending and incumbents' share of the vote. For challengers, on the other hand, spending contributes to the volume of the campaign, placing information about the candidates before the voters; as a consequence, it is more likely to have an impact.

Jacobson concludes his latest (1985) survey by noting that although 'the evidence is overwhelming that the challenger's level of spending has a strong impact on the vote, whereas that of the incumbent has virtually no impact at all' (p. 55), this may be a result of technical problems in circumventing the simultaneity issue — likely winners are more able to raise money, so that the spending reflects that likelihood rather than influencing the result. Nevertheless, Jacobson believes that the conclusions probably are valid, so that in terms of campaign finance policy, 'restrictions on campaign money will have the general effect of hurting challengers'. For them, spending is necessary in order to ensure that the electorate is well-informed about their candidacy and what they stand for; for incumbents, spending is sometimes necessary to counter a vigorous campaign on its contents, but is not needed for providing information. In many campaigns incumbents do not *need* to spend, but they cannot gamble on that, since the need may not be evaluated in advance. In any case, they find it relatively easy to raise money, and so most of them run an expensive campaign.

Virtually all of Jacobson's analyses refer to contests involving an incumbent candidate. What of the so-called 'open seats', in which both parties are running non-incumbents? On average, expenditure on these is much greater, especially in the House elections (Jacobson, 1980, p. 55), because both parties perceive the possibility of victory. For them, Jacobson used a simple model,

$$DV = f(DE, RE, DPS) \tag{2.4}$$

where
DV is the Democrat share of the vote;
DE and RE are Democrat and Republican expenditure, respectively; and
DPS is Democrat party strength in the district, indexed by its share of the vote at the previous election.
In analyses of six elections (1972 to 1982, inclusive), RE had a significant, negative impact in each case, whereas DE had a significant positive impact in only three. Jacobson (1980) suggests that the greater influence of RE reflects the Republican party's general disadvantage with the electorate (in terms of party identifiers), that has to be overcome by information provided through campaign spending; for Democrats, on the other hand, spending to provide information is not so greatly needed, and only influences the content of the contest on some occasions.

In summary

What these North American studies have shown can be summarised as follows:
(1) Challengers almost invariably benefit from expenditure, in terms of votes won, whereas incumbents rarely do.
(2) Incumbents can usually raise large campaign funds, whereas for challengers raising money is more difficult; those challengers most likely to win are best able to attract contributions.
(3) Since incumbency is a major advantage in American elections, in terms of presenting information about the candidate to the electorate, campaign spending will have little informational impact — hence its lack of effect on the outcome. For non-incumbents, however, other sources of information provision are much weaker in general, so that campaign spending is necessary to establish their presence before the electorate. Hence, although non-incumbents with a good chance of winning are better able to raise funds than 'no hopers', nevertheless those funds are likely to have an impact. The

possibility of victory has to be built-upon by seeding the local environment with relevant information.

These findings do not translate to the British experience, because of the differences between the two political systems. Nevertheless, they provide a useful background against which a study of British electoral behaviour can be set.

SPENDING AND VOTES IN BRITAIN

Although, as pointed out in the previous chapter, Parliamentary candidates have been required to report on their expenditure since 1885, and these data have been published, relatively little academic attention has been paid to the amount spent on the local campaigns, let alone on its electoral impact. (A recent book on turnout — Mughan, 1986 — makes no reference at all to the impact of spending.) In general, British psephologists believe that: (1) the data are unreliable, since candidates and their agents are able to obtain substantial price reductions from sympathetic suppliers (see Gordon and Whiteley, 1980); and (2) that British campaigns focus on parties nationally, with local activities having very little influence on the result. (With regard to the first point, presumably the data are only unreliable if candidates — or more properly, their agents — vary in their ability to obtain reductions.)

This set of attitudes is well exemplified in the authoritative Nuffield College series of books on British general elections which have appeared since 1945. The first volume in the series (McCallum and Readman, 1947) had a substantial section on the costs of constituency campaigns (pp. 39–41), and a three-page appendix listing the amounts spent; this concluded that 'The party analysis shows that the Conservatives made most use of funds but it was unable to save them from heavy defeat' (p. 297). Similarly, the 1950 volume reported in detail on the changes in the law relating to expenses (Nicholas, 1951, pp. 14–21). In analysing the amounts spent, several of the arguments developed here are present with regard to the pattern of spending. Thus,

> Parties seem to put more into seats they win than seats they lose
> . . . In the 20 seats where the fewest votes divided the major
> parties both sides, as might be expected, spent most heavily. The
> Conservatives virtually spent the permitted maximum while
> Labour spent 88% of it (pp. 18–19).

There was little analysis of the impact of that spending, however, except with regard to the Liberals:

> Over the country as a whole . . . they spent only 55% of the permitted amount. But in the 20 seats where they fared the best the average expenditure rose from £459 to £738, or 92% of the legal limit. This suggests that they may have been fairly successful in concentrating their expenditure where it would do most good, though it may merely reflect the financial self-sufficiency (or insufficiency) of strong (or weak) constituency organisations (p. 19).

The link between strong organisations, money-raising, and electoral outcome was not explored. The chapter on 'The campaign in the constituencies' made no reference at all to expenses.

The attitude set in the initial Nuffield studies, therefore, was that constituency expenses had no impact on the result, and this is exemplified in the more recent volumes. Writing of the 1970 election, for example, Butler and Pinto-Duschinsky (1971) noted that Conservative candidates on average spent more than Labour contestants (79 and 68 per cent of the legal maximum respectively), who in turn spent more than Liberal candidates (an average of 41 per cent). They admitted that more money was spent where the potential returns were greatest — 'the parties spent near to the hilt in marginal seats' — but claimed, without any analysis, that 'it would be hard to show any correlation between expenditure and success' (p. 334). Four years later, the report on the amount spent noted simply that 'The Labour and Conservative parties do . . . spend near to the limit in the marginal seats' (Butler and Kavanagh, 1974, p.240). Regarding the second election in that year, they added (citing Hill, 1974) that 'it is not apparent that greater local expenditure wins elections: recent research suggests that there is no relationship between the amount of money a party spends at an election and the change in its share of the vote' (Butler and Kavanagh, 1975, p. 244). They repeated this conclusion for the next election, five years later, arguing, first, that spending was high in marginal seats, and secondly, that anecdotal evidence counters any general trend. Thus (Butler and Kavanagh, 1980, p. 316):

> The Liberals were at no disadvantage [in amount spent] where it mattered most: in their 30 best constituencies they spent 91% of the maximum, while the 13 MPs elected in October 1974 spent

94%. In the 62 most marginal of Conservative/Labour contests (under 5% majorities in October 1974) Conservatives spent 91% of the maximum, and Labour 87%. Money is not essential to success. George Younger spent only £682 defending his safe Conservative seat in Ayr while Ernest Armstrong spent only £666 in his even-safer stronghold of North-West Durham. Jocelyn Cadbury gained Northfield for the Conservatives on £1472 (50% of the maximum) and David Samuel failed by only 618 votes to win Mitcham and Morden on an outlay of £1242.

Finally, for the 1983 election (Butler and Kavanagh, 1984) they similarly wrote that

it was notable that in the seats they won, the Conservatives spent 85% of the maximum, Labour 74% and the Liberals 87%. However this was not so much a case of money buying victory as of money going to places where victory was probable or possible (p. 266).

There was no analysis to sustain this conclusion, yet it led them to write that

neither in the constituencies nor nationally was money enormously important in determining the election result. . . . In the constituencies a few strong personalities, attuned to local issues and skilled at exploiting media attention, managed to make a limited dent in the national trend. But the general election of 1983, even more than its predecessors, was a national battle rather than the sum of disparate local struggles (p. 267).

Throughout, Butler and his associates equate description (where money is spent) with analysis (the impact of this expenditure).

The attitudes towards the impact of constituency spending displayed by Butler and his co-workers appear to derive from the conclusion to Kavanagh's (1970) study of *Constituency electioneering in Britain*. At several places in that book, he notes that the constraints on spending during the campaign period are severe, so much so that they have influenced the nature of the campaigning:

The imposition of this financial straightjacket has encouraged candidates to rely on well-established and inexpensive techniques and discouraged them from attempting other, and perhaps more expensive tactics (p. 15) . . .

. . . only in the Labour party is the question of the availability
of funds . . . still important; campaigning in the twentieth
century has ceased to be a money-spending activity for the
candidates and become a money-earning one (p. 30).

With regard to other activities, Kavanagh argued that analyses of
campaigns are characterised by the absence of 'hard' data (p. 37),
though he provides none for spending (despite quoting, without
comment, the response of many candidates to his inquiries that the
spending limit should be raised). Instead, he argues that campaigns
in Britain, where politics are effectively 'nationalised', have little
impact on the result. They do, he suggests, help to legitimise elections through the voter-candidate contact which occurs; in addition,
candidates '. . . have to act on the assumption that their activities are
important and *might* be decisive in winning votes. Such a belief is
sustained by the odd freak election result and the prized voter who
has been won over' (p. 111), but in general they are 'something of
a confidence trick for candidates'.

Early impact studies

The first study of the impact of campaign spending in British elections used no formal methods of analysis. Pennock (1932) noted an
average expenditure per candidate in a general election campaign
period in the late 1920s of about £720, plus an annual expenditure
per constituency (on the employment of an agent in particular) of
about £400. The total annual expenditure (also including local
politics) he estimated at £1.3m. (about £17m in 1986 costs), and he
asks:

is this sum really large considering the uses to which it is put?
. . . how can democratic government operate without elections,
and how can elections be carried on without organizations to fight
them, and how can elections be fought without money? Furthermore, is it not vital to a democracy to educate the public in
political matters, and how can this be done, at least under present
conditions, without political parties? (p. 20).

The parties spent very different amounts of money at the time,
however; Pennock estimates (p. 23) annual expenditure by the

36

Conservative party at £600,000, by the Liberals at £400,000, and by Labour at £300,000. If this reflects different levels of organisation, and hence of 'education', then presumably there should be differential benefits accordingly. Pennock suggests that expenditure goes on three types of activity. The first is what he calls *nursing* — 'the efforts, both financial and other, which a member of Parliament or a prospective member makes in cultivating the favour of his constituency' (p. 90). This could involve expenditure in the constituency — contributing to charities, paying the costs of local organisations etc. — by the individual; increasingly it involved the 'local ombudsman' role of dealing with constituents' complaints. The others are *education* and *propaganda*. The former is the long-term activity (akin to political socialisation — p. 5), designed to keep the electorate informed about political issues and the party's response to them. The latter — political evaluation (p. 6) — dominated the election campaign. ('When elections come along, the level of the educative work falls and the type of party propaganda descends to a somewhat lower level' — p. 97.) The goal is to inform, not to corrupt, and Pennock argues that

> one is not likely to see any candidate or any party neglecting to spend as much money on the election as the law allows or as much as he (or it) can collect. It would not be safe to take a chance, and politicians realise that even with the most careful work before the election, it is quite necessary to keep their forces in line during an election campaign. You may not be able to attract many new voters, but you must at least keep your regular ones in the harness (p. 93).

Does it work in that way? Does expenditure on propaganda keep the faithful in the harness and also win new support? Pennock offers no formal analysis, but writes as follows.

> A glance at the returns of election expenses will demonstrate that in Scotland [over the 1922, 1923, 1924 and 1929 elections] the best-financed candidates are most often unsuccessful; that in the rural areas the largest funds seem to be spent by the successful candidates, and that in the urban areas sometimes the best-financed candidate wins and sometimes the victory goes to his poorer opponent (p. 111).

As noted above, the three parties differed very substantially in what they spent (and presumably were able to raise) on this form of political advertising. However, Pennock seems to believe that this is unimportant, for

> if elections went according to the size of the money-bags of the respective parties, the results would be entirely different from those actually secured. We see consequently that money does not swing elections, even though all parties use strenuous efforts to collect as much of it as they possibly can (p. 111).

However, the parties, at least those which spent least, were not certain, and wished to insure against the possibility that campaign spending was effective, and hence they would not be disadvantaged. In the House of Commons debate on the Franchise Act in 1928, both the Labour and Liberal parties were in favour of reducing the maximum that could be spent, but they could not outvote the Conservative party on this issue. In 1949, when Labour was in power, such a reduction was introduced, so that a borough constituency with 50,000 electors had its maximum reduced from £1040 to £762. (The figures are taken from Pinto-Duschinsky, 1981a, p. 250.)

Recent analysis

Very little statistical analysis of the pattern and impact of spending has been undertaken. The first was a short paper by A. Taylor (1972) who argued that constituency spending was a reasonable index of party organisation there. He correlated change in spending between the 1966 and 1970 general elections with change in vote share, using only four categories for the former variable and average vote share changes for all constituencies in the category as the dependent variable. He concluded that in general, greater spending brought a greater return in terms of votes, but, as Hill (1974) points out, rank correlations based on only four observations are not particularly meaningful. Using continuous data for the same pair of elections, he reported that the only significant correlation related to the positive impact of Liberal spending on the Liberal vote share. He concluded that there were neither theoretical (because of the continuity factor in British politics) nor empirical grounds for 'positing a simple causal relationship between election expenditure and votes' (p. 217), but he did suggest that the impact of spending should be analysed with the previous election result held constant, thereby controlling for the continuity factor.

Apart from these two papers, the only other analyses have been by the present author. These began by testing the models of Palda and Welch, reviewed above. Later studies — in particular a sequence presenting various forms of analysis of the 1983 general election — have evaluated Jacobson's work in the British (largely English) context.

The production function model (p. 24) was explicitly tested by Johnston (1979a) on the results of the October 1974 general election in England. The model initially formulated was

$$v_{ij} = f(E_{cj}, E_{lj}, E_{bj}, I_{ij})$$ (2.5)

where

v_{ij} is the percentage of the votes won by party i in constituency j;
E_{cj}, E_{lj}, E_{bj} are the percentages of the allowed maximum spent, respectively, by Conservative, Labour and Liberal in constituency j; and
I_{ij} is a dummy variable, coded 1 if party i won in constituency j in February 1974.

(The first four variables were expressed in logarithmic form, to allow for potential diminishing returns from additional expenditure.) The results were encouraging, with high values of R^2 and with all but one of the regression coefficients (that relating Liberal spending to Conservative votes) being statistically significant. As anticipated from the production function model, holding incumbency constant (i.e. possession of the seat), the more that a party spent on advertising, the more votes it received, whereas the more its opponents spent the less votes it received. There was one exception, however: the more that the Conservative party spent, the more votes that the Liberal party received. The implication is that Conservative information woos voters away from Labour, but that some of them vote Liberal rather than Conservative. (Studies of voting behaviour in the 1970s — e.g. Himmelweit et al., 1985 — show that the Liberal party was a 'half-way house' for those deserting one of the two main parties but unable to shift all the way to the other.)

The model was then extended to incorporate the simultaneity problem (p. 29), introduced by Jacobson and used by Butler and Kavanagh to suggest why expenditure has no impact on election results in Britain. A further function was specified as

$$E_{ij} = f(I_{ij}, M_{ij}, TV_j)$$ (2.6)

39

where

E_{ij} is the spending by party i in constituency j as a percentage of the maximum allowed;

I_{ij} is the incumbency variable defined above;

M_{ij} is a measure of the marginality of the seat (the larger the value, the safer the seat); and

TV_j is the total number of voters in constituency j.

Thus, it was suggested, parties should spend more in seats that they already held; in more marginal seats; and in the larger constituencies. The results did not indicate close fits, so that although in general more was spent defending what was already held and in fighting the marginal seats, the raising and spending of money locally was not particularly 'rational'. A final test, that the change in a party's performance between February and October 1974 was related to the change in its spending between the two, produced no evidence in support of the model.

An earlier study of Scotland (Johnston, 1977) had tested a similar sequence of hypotheses for the February-October 1974 election sequence, with comparable results. Each of the four parties spent more in October in the seats that it won in February than in those that it lost, and spent more, the greater the marginality of the seat (i.e. its percentage of the vote in February relative to that of the winner if it lost, and relative to that of the second-placed if it won). Change in the pattern of expenditure between February and October had very little impact on the result in the latter contest, however. Further analyses of both Scotland and Wales extended this work (Johnston, 1979b). At both elections in 1984, the model derived by Welch (1974) was found valid; comparison of the regression elections showed that in Scotland, the weaker the party overall, the steeper the relationship between marginality and spending (Taylor and Johnston, 1979, p. 311).

A final test of the Welch/Palda model was concerned with the 1970 general election result in London and with the 1979 European Assembly Election in England (Johnston, 1983a). For London, the results were similar to those discussed above. For the European Assembly election, only Liberal voting had any impact on turnout, and only Liberal spending had any significant impact on the vote share. The conclusion was that 'The analysis of general election voting . . . shows no strong support for any influence of spending on changes on the pattern of votes, and the analysis of the first European election shows no clear relationships between spending and votes obtained' (p. 125).

Given such findings — added to the writings of Butler, Kavanagh, and Pinto-Duschinsky — the likelihood of any different conclusions emanating from a study of the 1983 election seemed remote. However, the presence of the Alliance introduced a novel element that could have led to new patterns of spending and impacts, and experimentation with other models suggested problems with the specification used in earlier studies (as Gordon and Whiteley, 1980, had suggested). The initial analyses involved addition of spending levels to a study of spatial variations in the flow-of-the-vote, as estimated by entropy-maximising procedures (Johnston, 1985a). The conclusion was that 'Where a party spent more, so more of its 1979 supporters remain loyal, *ceteris paribus*, and fewer either transferred their allegiance to the other parties or failed to vote' (p. 273). This was confirmed in a later, more complete investigation of the pattern of net flows, with Liberal spending having a greater impact than SDP spending (Johnston, 1986f). The differences between the two Alliance parties (stressed in Johnston, 1985e) were crucial, but none of the tests in this sequence of papers also distinguished, following Jacobson, between the influence of spending by incumbents and spending by challengers (see also Johnston, 1985f). Once this had been done (Johnston, 1986g), the greater importance of spending by challengers was appreciated. Finally, the model was refined to remove some statistical problems (Johnston, 1986h) forming the basis of that presented in Chapter 3 and evaluated here for a sequence of elections, and estimates made of the impact of spending variations on the number of seats won by each party (Johnston, 1986e).

IN SUMMARY

Campaign spending has been much more extensively studied in the United States than in the United Kingdom, in part because of the much greater sums expended in the former. Analyses there have shown that the amount spent has significantly influenced many results, especially the amount spent by challengers. Most writers on British elections deny the existence of any link between spending and vote-winning, but draw their conclusions without undertaking detailed (even simple) statistical analyses. However, a sequence of papers applying the American findings to the British situation casts considerable doubt on those conclusions. They provide the basis for a model to be developed in Chapter 3 and evaluated in the rest of this study.

3

Towards a Model

Studies of the impact of campaign spending in the United States, reviewed in the previous chapter, suggested that for non-incumbents, the more that they spent on providing information for the electorate, the more votes they obtained. Early studies in Britain provided little evidence of a similar relationship between spending and votes won, but later work on the 1983 general election produced results consistent with the American findings. The present chapter builds on those findings, developing a model of the spending:votes relationship, to be tested in England; other tests in Chapter 6 use particular extensions of the model developed here.

PRELIMINARIES

Three features of English elections are relevant to a study of the impact of spending, and should be incorporated into any model.

Continuity

As in most electoral systems, a major characteristic of voting behaviour in England is its continuity. At the individual level, this is shown by the percentage of the electorate who remain loyal to a party over a sequence of elections. Until the 1970s, this was very substantial: Butler and Stokes (1974, p. 273), for example, show that 68 per cent of those who preferred Conservative in 1959 did so again in 1970, as did 56 per cent of those who voted Labour in 1959. The 1970s, according to Sarlvik and Crewe (1983), was a decade of dealignment: thus, whereas 74 per cent remained loyal (including abstaining twice) between 1964 and 1966, only 69 per cent did so between the two 1974 elections and 62 per cent between 1974 and 1979 (Crewe, 1985a, p. 110). Even then, loyalty remained the dominant characteristic.

At the aggregate level, the evidence of continuity is even greater. Analyses of data at constituency scale suggest an unchanging geography of voting since the 1920s (Johnston, 1983a, 1987a; Taylor, 1982). The reason for this, according to Johnston, O'Neill and Taylor (1986), is the role of contextual influences in political socialisation (see also Johnston, 1986a, 1986b). People learn political attitudes in places, and their party loyalties are thus in part community-based. Hence, individual continuity is exaggerated at the aggregate scale; the political context of places is relatively permanent.

In recent decades there have been some changes in the geography of voting in England, as portrayed by Curtice and Steed (1982). The reasons for this are far from clear (for a recent discussion, see Savage, 1987). They appear to be permanent, and only partly related to changes in local milieux (as in the spatial differentiation in the growth of unemployment between 1979 and 1983 — Johnston, 1983b, 1985a). Furthermore, they are almost certainly not produced by campaign spending, the impact of which is more likely to be in short-term shifts.

Given this continuity in electoral behaviour, the impact of campaign spending is likely to be small, producing minor variations around the general continuity. Thus, any analysis of the impact of spending at one election must take account of the result at the preceding contest. By taking account of continuity in this way (i.e. by regressing the result at the second election on that at the first), two potential problems are avoided. First, there is no need to consider the wide range of variables underlying that continuity — the various aspects of the social and economic geography of England which are related to its electoral geography (Johnston, 1985a). Some short-term changes are also related to those variables but, as indicated here, they are largely unrelated to spending. Secondly, there is no need to consider the change in the level of spending from one election to the next (as done by Taylor, 1972, and Johnston, 1979a, 1979b). It could be argued that the result at the first in a pair of elections reflects, in part, the spending then incurred, so that any deviation from it at the second election should reflect not the level of spending, but changes in the level of spending. What is argued here, however, is that because of continuity, the pattern of voting at one election is very closely related to the pattern at the next. Any deviations from that pattern should be the consequence either of major changes in the nature of a constituency (social, economic, or political) or of attempts by parties to influence the election result there through the processes of political evaluation. Those attempts are time-specific, and are the focus here.

Two types of incumbency

In the United States, as a great number of studies has shown, incumbency is a major influence on election results. Candidates in possession of a seat have a very strong initial advantage — a consequence, it seems, of their ability to obtain media coverage, to promote themselves, via unlimited free access to the postal system, to the electorate at relatively little cost, to raise campaign funds from sponsors who hope to obtain benefits, and to present themselves as able to win pork-barrel benefits for their constituents (Johnston, 1980a). The link to party may be important, both as a source of funds and in terms of the general partisan attitudes. However, it is the individual candidate's incumbency that is paramount, so much so that most analyses classify contests for seats not being fought by the incumbent as 'open', even though one party may have won by a landslide at the previous election.

In Britain, on the other hand, ties between electorate and party are very much stronger than those between electorate and candidate. The British Parliamentary system involves MPs acting out a variety of roles. Bogdanor (1985) identifies four: (1) as a representative of *constituency* interests; (2) as a representative of *partisan* interests; (3) as a protector of *sectoral* interests; and (4) as a legislator promoting certain *policy* options. He claims that the second is dominant, so that 'the parliamentarian will see himself . . . less as a representative of a constituency than as representative of a party point of view' (p. 5). Hence, party solidarity in the House of Commons is great, and voting against one's party is a matter of considerable significance, treated to a great deal of media comment (see Norton, 1978).

A consequence of the dominance of party machines in Parliament should be the relative national anonymity of most individual members — except those on the front benches and the few who obtain national reputations, often notoriety. This would lead to the dominance of party over candidate in election campaigns. However, in their own constituencies many MPs do a great deal of work meeting the electorate and handling their complaints about the implementation of a vast range of official policies, and most MPs are able to obtain substantial coverage in the local press. Despite this, surveys usually show a large percentage of electors who know very little about their local MP, even his or her name, and a very small number who have had any personal contact with their representative (data taken from Crewe, 1985b).

Despite the strength of the party dominance, incumbent MPs, especially those who are diligent in their constituency activities, should gain some electoral reward from their position in the local public eye and the services that they render to their constituents. The size of that reward is difficult to estimate; as Crewe (1985b, p. 58) puts it, 'disentangling the MP's individual appeal from his party's, and from other influences peculiar to the constituency, makes precise estimates notoriously tricky'. Both he and Berrington (1985) draw on the work of Curtice and Steed (1980), who looked at election data, and on Cain's (1983) study of MPs' constituency activity to suggest estimates of the possible importance of incumbency. Berrington (1985, p. 34) claims that 'There are signs, too, that a record as a good "constituency Member" helps the MP to withstand a swing against his party or to do better, if the tide is flowing with his party, than his colleagues'. Crewe is more circumspect, arguing, with Curtice and Steed, that the personal vote is likely to be established the first time that the incumbent seeks re-election, and that there is no continuous building-up of support over a sequence of elections. Hence his conclusion that 'For the vast majority of MPs good works do not save — and waywardness does not damn' (p. 58).

These discussions of personal votes focus very much on the incumbent MPs for the two parties — Conservative and Labour — which have dominated British politics in recent decades. What of other candidates, the non-MPs and the standard-bearers for the smaller parties (the Liberals, the nationalist parties and, in 1983, the SDP) as well as the independents? For these, their party may gain very little national recognition, and relatively few voters are permanently aligned with them. It is for the individual candidates to build local support through intensive activity. Curtice and Steed (1984) noted that this is almost certainly the case with the Liberal party, and suggested that many of the Alliance candidates in 1983 who had been elected in 1979 as either Labour or Conservative also retained substantial support that can only be interpreted as a personal vote. Nevertheless, Crewe (1985b, p. 63) claims that

45

The two lessons learned by the SDP defectors underline the evidence of this paper: that the local MP in Britain does not have the resources or opportunity to build up a personal power base in the constituency; and that party loyalty in the electorate, although weakening, is still a dominant force.

The implication, therefore, is that party incumbency is much more important than candidate incumbency as an influence in British election results. The party holding a given seat has a substantial advantage. An incumbent MP might have a further advantage of a few hundred votes. Whether a candidate who fought the seat before but lost has any advantage has not been fully analysed, though one who 'nurses' the constituency diligently throughout the Parliament may build some personal support there.

Context

The literature on voting in Britain is dominated by a compositional theory of behaviour (albeit implicit in most cases; a clear, explicit statement is that of Dunleavy and Husbands, 1985). Individuals are allocated to a socio-demographic category — usually one in which socio-economic criteria, such as occupation, are dominant — and their political attitudes and voting behaviour are assumed to follow. This assumes also that the milieux within which people learn the meaning of such membership are all the same, so that, for example, to be a manual worker living in council housing means the same in any part of the country: such an assumption is made explicit by Bogdanor (1983) as quoted above (p. 10).

However, milieux do vary substantially. Places can be characterised according to three criteria: (1) their position in the division of labour — what is done there; (2) the structure of social relations there; and (3) the local institutional framework. These represent the economic, social, and political components of life (see Massey, 1984; Johnston, 1986c). Because two places are similar on one of the criteria, it must not be assumed that they are similar on the other two as well, so that, for example, social relations and political institutions may vary among coal-mining communities, with consequences for political behaviour (see Johnston, 1986c).

Milieu is important to the development of political attitudes and the resultant pattern of voting behaviour not only because people learn about politics in places, but also because political parties are active in places, seeking to influence that learning; by influencing the short-term processes of evaluation, they also hope to affect the longer-term processes of political socialisation. Parties are not national organisations alone, canvassing for votes at election time through the mass media: they are also local institutions, and crucial agents of political socialisation, pursuing place-based strategies in the search for commitment and votes (Taylor, 1985a; Johnston, O'Neill and Taylor, 1986).

Most approaches to the analysis of voting behaviour which incorporate the contextual effect do so only partially through the inference of what is widely known as the neighbourhood effect. A number of studies — notably that of Miller (1977) — have shown that the propensity for people in particular socio-economic categories to vote for a certain party varies according to the neighbourhood context. Given the underlying class basis of British voting, what this shows is that the greater the dominance of one class in an area, the greater the likelihood that members of all classes will vote for the party favoured by the numerically dominant group. The rationale for this is unclear, as Dunleavy (1979) has trenchantly argued; Taylor (1985a) has indicated how it can be accounted for in terms of a theory of voting that incorporates contextual as well as compositional variables (see also Johnston, 1986a, 1986b).

Contextual effects should be relatively permanent, of course, given the relative stability of local milieux, and so for certain types of electoral analysis can be incorporated with the continuity effects discussed above. However, changing patterns of partisan allegiance may also possess a distinct geography, reflecting changes in the local milieux (on all three of the components that define a place) and in local political activity. Those changes in the milieux may be permanent, as with economic and social restructuring, or they may represent temporary influences, such as high levels of unemployment. In England between 1979 and 1983, both were apparently present. Analyses of the flow-of-the-vote (Johnston, 1985a) showed a continued spatial polarisation — Labour did relatively well in the northern region and the Alliance relatively badly; the Conservative party performed relatively badly in the inner cities. In addition there were above-average shifts to non-voting in areas with either high levels of unemployment or large percentages employed in the energy industries, and Labour remained relatively strong in those areas. Thus, although overall the contextual effects identified by analyses of voting at a single election are of little importance in studies of change over time, because they are incorporated in the continuity factor, it is possible that local factors can influence the trends and produce deviations from the general pattern. One such local factor, and the only one focused on here, is party campaign spending.

STRUCTURING THE MODEL

The focus of the analyses to be reported in this study is on the role of constituency campaign-spending as an influence on election results, but that influence is to be studied within a model that incorporates other influences. Since the concern of a study of campaign spending is that of changing patterns of voting, this suggests the following set of hypotheses.

H_1 A party's share of the vote at any election in each constituency is closely related to its share at the previous election there. (*The continuity hypothesis.*)

H_2 A party's ability to raise funds for the campaign in a constituency is much greater if it holds the seat than if it does not. (*The party fund-raising hypothesis.*)

H_3 A party's ability to raise funds for the campaign in a constituency is greater if its candidate was also the candidate at the previous election than it is if a new candidate is being fielded. (*The candidate fund-raising hypothesis.*)

H_4 A party's level of spending in a constituency reflects the constituency's marginality at the last election. (*The marginality hypothesis.*)

H_5 A party's share of the vote at any election in a constituency is greater if its candidate was also the candidate at the previous election than if a new candidate is being fielded. (*The candidate incumbency hypothesis.*)

H_6 A party's share of the vote at any election in a constituency is a function of the amount that it spends on the campaign. (*The party-spending hypothesis.*)

The first of these hypotheses, H_1, represents the continuity factor so important in British voting behaviour at both individual and aggregate scales; it provides the baseline against which changes induced by spending are to be measured. The next three (H_2, H_3, and H_4) relate to the pattern of spending and suggest that parties will spend more: (1) if they hold a seat than they will if they are providing a challenger; (2) if their candidate contested the constituency at the previous election: and (3) the greater the chance of winning/losing the seat because of its marginality. They clearly suggest that the pattern of spending reflects aspects of the local milieux, particularly the anticipated consequences of not working hard there to influence the processes of political evaluation (H_4). The last two hypotheses (H_5 and H_6) refer to the impact of that campaigning, arguing that both incumbent candidates and candidates who spend more benefit more, in terms of votes. Together, these six hypotheses produce the model shown in Figure 3.1, where:

V1 is the percentage share of the vote won by the party at election 1;

V2 is the percentage share of the vote won by the party at election 2;

P INC is a dummy variable, 1 if the party holds the seat and 0 otherwise;

INC is a dummy variable, 1 if the party's candidate contested the seat at the previous election, and 0 otherwise;

M is the marginality of the seat; and

S is the amount spent by the party on the campaign.

Figure 3.1: The model of the pattern of spending and its impact: a first approximation

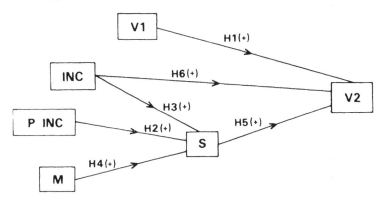

For each link, the relevant hypothesis is indicated, with a sign representing the direction of the expected relationship.

This model could be tested directly, for each party at each election. Several problems would arise, however.

(1) There is potentially strong collinearity among several of the variables. In particular, there will be a close relationship between V1 and P INC: if only two parties are involved, that relationship will be very close.

(2) The variables V1 and M will also be closely related — perfectly so if there are only two parties. The relationship will be U-shaped, if marginality is defined as follows: if the party holds the seat,

49

marginality is the difference between V1 and the vote share of the second-placed party at the first election; and if the party does not hold the seat, it is the difference between V1 and the vote share of the party that does hold it.

(3) The impact of incumbent candidates for other parties is not evaluated, nor is that of the spending by other parties.

(4) The dependent variables form a closed system. If there are three parties contesting each election — A, B, and C — then

$$V1_A + V1_B + V1_C = 100 \qquad (3.1)$$

and

$$V2_A + V2_B + V2_C = 100 \qquad (3.2)$$

The three tests of the model should be complementary, but difficulties will be encountered in evaluating the impact of the various influences on the result.

To avoid these problems, the following strategies have been evolved (as in Johnston, 1986h);

(1) The set of constituencies will be divided according to which party won each at the first election in the pair being studied. This avoids the collinearity problems involving V1, P INC and M and the U-shaped relationship between V1 and M, and allows direct testing of the impact of challengers' spending.

(2) Spending by each party is incorporated.

(3) Whether each party has an incumbent candidate is incorporated (i.e. a candidate is classified as an incumbent if he/she stands for the same seat at two successive elections, irrespective of the result of the first).

(4) The impact of campaign spending on vote share is evaluated directly by using the ratio between the vote share of the two parties concerned instead of V1 and V2, both as an independent variable — referring to the first election of a pair being studied — and as a dependent variable — referring to the second election of the pair. For a two-party situation, this gives the model of Figure 3.2, referring to seats held by party A, where:

[A/B]1 is the ratio between the share of the vote won by party A and that won by party B, at election 1 — it will always be greater than 1.0;

[A/B]2 is the ratio between the share of the vote won by party A and that won by party B, at election 2;

S_A is the spending by party A at election 2;
S_B is the spending by party B at election 2;
INC_A is a dummy variable, 1 if the candidate for party A fought election 1 as well as election 2, and 0 otherwise;
INC_B is a dummy variable, 1 if the candidate for party B fought election 1 as well as election 2, and 0 otherwise.

Figure 3.2: The basic model. The number alongside each link refers to its listing in the text

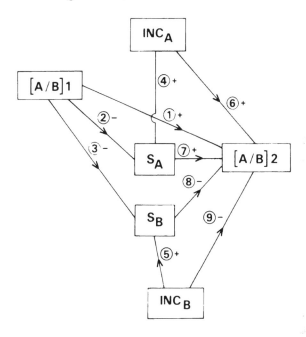

Nine paths are shown in the model. The expected relationships for the relevant partial regression coefficients are as follows.

(1) Positive; a high ratio at one election should be followed by a high ratio at the next. (*The continuity hypothesis.*)

(2) Negative; a high ratio will mean low marginality, and less need to spend. (*The marginality hypothesis.*)

(3) Negative; as for link 2.

(4) Positive; an incumbent candidate (and MP, since A holds the seat) should be better able to raise campaign funds. (*The candidate fund-raising hypothesis.*)

(5) Positive; as for link 4 (except that the candidate will not be an MP).

(6) Positive; an incumbent candidate for A should be better able to

win votes, and so increase the A/B ratio. (*The candidate incumbency hypothesis.*)

(7) Positive; the more that party A spends, *ceteris paribus*, the greater its share of the vote should be and so the higher the [A/B]2 ratio. (*The party-spending hypothesis.*)

(8) Negative; the more that party B spends, *ceteris paribus*, the smaller A's share of the vote should be and so the lower the [A/B]2 ratio.

(9) Negative; if party B fields an incumbent candidate, it should be able to lower the [A/B] ratio, *ceteris paribus*.

This tests all of the six hypotheses set out above except for the party fund-raising hypothesis (H_2). This can be evaluated by a simple difference-of-means test between the seats held by A and those held by B.

All of these tests assume a linear relationship between the relevant pairs of variables. As indicated in the previous chapter, several of the American studies identified curvilinear relationships between spending and vote-winning, suggesting that additional spending above a certain level produced diminishing returns (both positive and negative). For all of the analyses reported here, plots of the partial relationships between spending and votes were produced: examination showed no clear evidence (and certainly no general evidence) of curvilinear relationships, and so the linear format was retained.

Fitting the model of Figure 3.2 involves three multiple regression equations, as follows:

$$S_A = f([A/B]1, INC_A) \tag{3.3}$$

$$S_B = f([A/B]1, INC_B) \tag{3.4}$$

$$[A/B]2 = f([A/B]1, INC_A, INC_B, S_A, S_B) \tag{3.5}$$

In the first two, the impact of marginality and incumbency on the levels of spending is being assessed, so that for each, the partial regression coefficients for the links concerned — 2 and 4 in the first, and 3 and 5 in the second — reflect the independent effects of the two variables. In the third equation, links 1, 6, 7, 8 and 9 are all represented by partial regression coefficients, so that, for example, the impact of spending independent of continuity is being assessed and vice versa. Continuity is indexed by the relative strength of the partial regression coefficient for [A/B]1; the size of that coefficient indicates the average experience of the two parties between the elections (a coefficient of greater than 1.0 would indicate a swing towards the first-named party — i.e. the party holding the seat — whereas a coefficient of less that 1.0 would indicate a swing away from it); and the partial coefficients with the spending variables indicate the relationship between spending and deviations from the continuity factor, holding the average swing constant.

OPERATIONAL FORMS

The basic model of Figure 3.2 provides the structure for all of the analyses to be reported here. It is applied to a variety of electoral situations, however. For most of the period (1950–83) studied, British politics were dominated by two parties, Conservative and Labour. The Liberal party reached the nadir of its fortunes in the first decades of the period, and many constituencies were not contested by it in the 1950s and 1960s. This produced four types of constituency:
(1) Those fought by all three parties at both elections in the pair — the *Three-party* contests;
(2) Those fought by the Liberal party at the first election of the pair, but not the second — the *Liberal-exit* contests;
(3) Those fought by the Liberal party at the second election of the pair, but not the first — the *Liberal-entry* contests, and
(4) Those fought by Conservative and Labour only — the *Two-party* contests.
Separate models have been developed for each type.

Three-party contests

Figure 3.3 shows the models for these contests. In these, the variables are defined as follows:
CL1, CL2 are the ratios between the Conservative and Labour shares of the vote at the first and second elections of the pair, respectively;
LC1, LC2 are the ratios between the Labour and Conservative shares of the vote at the first and second elections, respectively;
CB1, CB2 are the ratios between the Conservative and Liberal shares of the vote at the first and second elections, respectively;
LB1, LB2 are the ratios between the Labour and Liberal shares of

Figure 3.3: The models for Three-party contests. For key to terms used see text

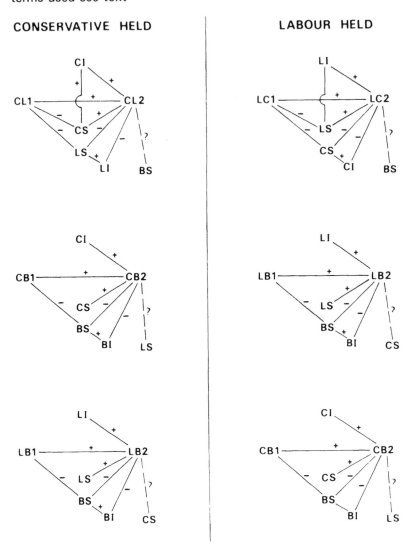

CONSERVATIVE HELD

LABOUR HELD

the vote at the first and second elections, respectively;

CS, LS, BS are the spending by Conservative, Labour and Liberal respectively at the second election, expressed as percentages of the allowed maximum because this varies by constituency (see below): and

CI, LI, and BI are dummy variables, 1 if Conservative, Labour and Liberal respectively fielded an incumbent candidate, and 0 otherwise.

The expected relationships are shown alongside the links in the models, of which there are three, representing the competition between each pair of parties for a share of the vote. For the competition between each of the two 'main parties' and the Liberals, no links are provided between either the ratio at the first election and candidate incumbency or the incumbent party with that party's spending; those relationships are in the first of the models (at the top of the diagram) only. Further, no direct link is provided for the relationship between the spending by the 'third party' in the model — i.e. that not involved in the vote ratios (Liberal in the first case, Labour in the second and Conservative in the third in Conservative-held seats) — because the direction of such a link could not be hypothesised *a priori*. That variable has been included in the regressions as an exploratory measure only, however; it is represented by the dashed lines (and is tested by two-tail rather than one-tail significance tests).

Two-party contests

The model for these is taken directly from Figure 3.2 (see Figure 3.4), using the same abbreviations as in Figure 3.3.

Liberal-exit contest

These two use the same model as Figure 3.2, so that Figure 3.4 refers to them as well as to Two-party contests. The presence of Liberal candidates at the first election of the pair is not considered, again because no firm *a priori* hypotheses could be derived. (It could

Figure 3.4: The models for Two-party and Liberal-exit contests

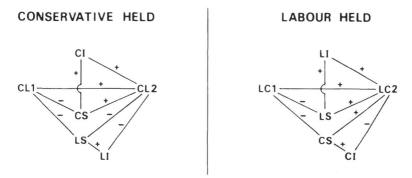

CONSERVATIVE HELD

LABOUR HELD

have been assumed that the greater the proportion of the votes won by Liberal candidates at the first contest, the greater the spending by the remaining two parties to win over those voters. However, relatively few voters remained loyal to the Liberal party between two elections during the 1950s and 1960s. No empirical evidence of such a relationship between prior Liberal vote and spending was found in early exploratory analyses.)

Liberal-entry contests

In these contests (Figure 3.5) the Liberal party is absent from the first election but present at the second. For the Conservative:Labour contests (the top pair in Figure 3.5), the models are those of Figures 3.2 and 3.4, with the addition of BS as a possible influence in the outcome of the contest.

For the ratio between the vote share of the party holding the seat and that of the Liberal party (CB2 in the left-hand column of Figure 3.5, LB2 in the right-hand column), the success of the Liberal party is assumed to be a function of the spending and incumbency variables. together with the safety of the seat. The safer the seat, it is suggested, the greater the potential for Liberal success, since it could attract both voters from the losing party, who see their votes as wasted, and those from the winning party who wish to record a protest vote without damaging their party's chances of winning the seat. Hence the postulated negative relationships between CL1 and CB2 and between LC1 and LB2. As a corollary of this, the safer the seat, the greater the potential inroads the Liberal party should make into the position of the party currently in second place. Hence the expected negative relationships between CL1 and LB2 and between LC1 and CB2.

THE CHANGING IMPACT OF SPENDING

Pinto-Duschinsky's work, reviewed in Chapter 1, shows that the amount spent by the parties on local campaigns has declined substantially in real terms over the period being studied here. With the growing importance of the mass media in election campaigns, local campaigning has become less important, it is suggested, especially after the reform of 1969 which allowed candidates to indicate their party affiliation on the ballot paper. For the purposes of the present

Figure 3.5: The models for Liberal-entry contests

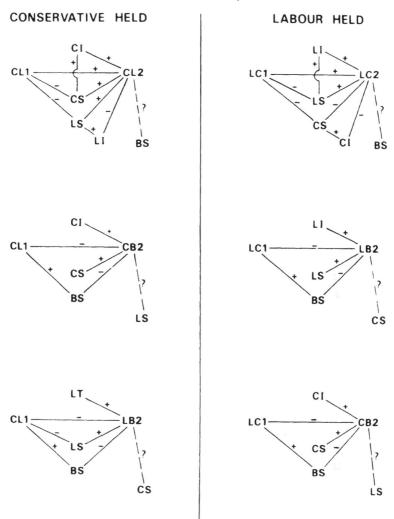

CONSERVATIVE HELD   LABOUR HELD

study, this suggests that local campaign spending should become less important and effective, leading to the following hypotheses.

H_7: Over time, candidates will be less concerned to raise money for the local campaign. This will be reflected by a decline in the significance of the relationships between incumbency and spending levels (e.g. the link between incumbency and spending in H_2 and the relationships between CI and CS and LI and LS in Figure 3.4).

H_8: Over time, parties will be less concerned with spending where it is most likely to matter. This will be reflected by a decline in the significance of the relationship between marginality and spending levels (e.g. the relationships between CL1 and CS and LS in the left-hand diagram of Figure 3.4).

H_9: Over time, spending will have less impact on the election results. This will be reflected by a decline in the significance of the relationship between spending and outcome (e.g. the relationships between CS and LS and CL2 in the left-hand diagram of Figure 3.4).

THREE STATISTICAL ISSUES

In testing these models three statistical issues arise with regard to the use of ordinary least squares regression models. First, there is the potential problem of collinearity, particularly among the spending variables. If, as the hypotheses suggest, parties spend more the more marginal the seat, then the levels of spending should be substantially intercorrelated. Examination of the correlation matrices for all of the analyses showed this not to be so (which raises doubts about either the rationality of the parties or their ability to respond to perceived marginality). The average correlation between Conservative and Labour spending was only 0.39, with a standard deviation of 0.17; only six of the values of r exceeded 0.5, and only two were so large that they indicated more than 50 per cent agreement (i.e. correlation of 0.707 or greater). For Conservative and Liberal spending, the average correlation was 0.24, with a standard deviation of 0.12; the largest individual r value was 0.48. And for Labour and Liberal spending the average was 0.09 (standard deviation 0.16) with the largest 0.52. Collinearity was clearly not a problem.

Secondly, there is the problem of homoscedasticity, which refers to equality of variances. If the distribution of one of the variables in a regression is skewed, heteroscedasticity is potentially present, which means that a relatively small number of outliers could have a major impact on the slope of the regression (see Jones, 1984, and Wrigley, 1984). In inferential statistics it is customary to correct for this by transforming the skewed variable, thereby ensuring efficient regression coefficients. However, the present analyses do not involve inferential work and although (as Chapter 4 indicates) many of the spending distributions are skewed, this reflects the tendency of incumbent parties especially to spend close to the maximum, with only a few outliers. Those outliers are crucial to the present exercise since they indicate relatively rare occurrences which give strong pointers to broader trends; transformations were not undertaken to reduce their important rarity value, therefore.

Finally, it must be restressed that all of the regression results predicted by the models refer to partial relationships — i.e. those between the independent specified and the dependent, holding all other independents constant. In assessing the importance of these relationships, the general test applied here is that they should be statistically significant at the 0.05 level. Given that the analyses do not involve samples, the inference to be drawn from such statements of statistical significance is that the relationship observed is of such strength (e.g. the goodness-of-fit of a partial regression line to a scatter of points) that it is very unlikely to have occurred by chance (i.e. the probability of it resulting from a random allocation of the observations is very small). The basic hypothesis of the study is that the volume of spending in the constituencies influences the vote there; a significant result is one that clearly substantiates that hypothesis. The importance of the hypothesised relationship relative to others is indicated in other ways (notably, by the relative size of the standardised partial regression — or beta — coefficients).

IN SUMMARY

A general model incorporating the impact of campaign spending into analyses of the pattern of voting in British elections at the constituency level has been presented here. This has been operationalised as a series of cross-sectional regression models which assess the influence of spending on the changing pattern of voting between a pair of elections. Particular hypotheses have been presented, and these will be evaluated for each pair of elections over the period 1950–83. (Some pairs are omitted, for reasons specified in the next chapter.) Longitudinal hypotheses have also been formulated, referring to secular changes in the influence of spending, and these will be evaluated via comparative study of the results of the cross-sectional analyses.

Presentation of the analyses and evaluation of the results occupies the next two chapters. These deal with elections in England only. Scotland and Wales have been excluded because of the development of nationalist parties there in the latter part of the study period; they are the subject of separate analyses reported in Chapter 6. Northern Ireland has been excluded because of its entirely separate party system, the small number of constituencies, and the strength of the partisan loyalty which suggest that campaign spending may be irrelevant. Chapter 4 is largely descriptive, setting out material on the level of spending and reporting on tests regarding levels of spending — between incumbent and challenger parties, for example. Chapter 5 presents the results of the regression analyses for England, and Chapter 6 evaluates the uses of the model in other contexts. The results are brought together in Chapter 7, which evaluates the findings both in the context of the hypotheses and with regard to possible policy implications.

4

The Pattern of Spending

This book is concerned with campaign spending by British political parties during a period when the real value of the sums expended declined by some 60 per cent. The elections covered run from 1951 to 1983, although the 1950 election is relevant to the analyis of 1951. The 1945 election was considered for inclusion but was excluded (thereby preventing detailed analyses of 1950) because of: (1) the unusual nature of that war-time election, the first since 1935; (2) the existence then of multi-member constituencies; (3) the boundary changes between 1945 and 1950; and (4) the changes in the legislation relating to campaign spending in 1949.

The models developed in the previous chapter are cross-sectional in that they refer to spending and votes at a single election, but the patterns of spending and votes then are set in the context of the voting at the immediately preceding election. Thus, every analysis refers to a pair of elections. This precluded the study of every general election since 1950 because of two major revisions of constituency boundaries (before the 1955 election and after the 1970 election; a third revision, in 1983, was not a problem because the 1979 results were reworked into the 1983 constituencies — BBC/ITN, 1983). Thus, the pairs of elections studied were: 1950–1951; 1955–1959; 1959–1964; 1964–1966; 1966–1970; 1974(February)–1974(October); 1974(October)–1979; and 1979–1983. They are referred to here by the second date in each pair. (Some constituencies were either unchanged or only slightly changed in both 1955 and 1970, so partial analyses of the 1955 and 1974 (February) elections would have been possible; these were not undertaken, however.)

The amount that can be spent during the official campaign period — i.e. between the declaration of the election and polling day — is

61

prescribed by the Representation of the People Act. Between 1949 and 1969 it was unchanged (Table 4.1); it was altered in the latter year and again in 1974, by amendments to the Representation of the People Act. A further amendment in 1978 also included the provision for the limit to be altered at the discretion of the Secretary of State, by an Order in Council, in line with the increase in the cost of living index (the term in the Act is 'change in the value of money').

Table 4.1: Spending maxima at each general election, and amount spent

| Election | Base Figure | Expenditure per Elector in | | Total spent |
		Borough Constituencies	County	
1951	£450	1.5d (0.63p)	2d (0.83p)	£0.743m
1959	£450	1.5d (0.63p)	2d (0.83p)	£0.869m
1964	£450	1.5d (0.63p)	2d (0.83p)	£1.018m
1966	£450	1.5d (0.63p)	2d (0.83p)	£0.882m
1970	£750	1.5d (0.63p)	2d (0.83p)	£1.091m
1974(Oct)	£1075	0.75p	1.0p	£1.733m
1979	£1750	1.5p	2.0p	£2.793m
1983	£2700	2.3p	3.1p	£4.881m

As will be seen from Table 4.1, the amount that can be spent varies between constituencies acording to: (1) their classification as either borough or county; and (2) the size of their electoral roll. The distinction between boroughs and counties predates the 1832 Reform Act, reflecting the different franchises between the two (Seymour, 1950, p. 10); more was traditionally spent on electioneering in the county constituencies (p. 406). When the constraints on spending were introduced in the Corrupt and Illegal Practices Act (1883), the differentiation was retained, on the assumption that electioneering is more expensive in low-density rural areas than in high-density urban environments. There are no strict rules for the allocation of constituencies to their categories, which is undertaken by the Boundary Commissions when they redistribute the seats; clearly, the implicit criterion is population density, but the House of Commons (Redistribution of Seats) Act (1949) refers only to the Commission stating 'as respects each constituency . . . whether they recommend that it should be a county constituency or a borough constituency' (Section 3). The commissioners make no mention of the criteria used in their first report (1954) under this Act.

Because of the variations in the amount that could be spent, it was necessary for all of the analyses here to use standardised data. (In 1955, for example, the maxima ranged in the English constituencies from £694 to £1034, with the largest nearly 60 per cent greater than the smallest; in 1983, the range was from £3770 to £5656, a difference of exactly 50 per cent.) This standardisation was done by expressing spending in each constituency as a percentage of the allowed maximum therein.

The amounts spent at each election remained relatively constant in actual figures during the period when the maxima were unchanged (Table 4.1). Since then they have increased, though less rapidly than the prescribed maxima. (In a borough constituency with 65,000 voters, the maximum rose by 528 per cent between 1966 and 1983, whereas the total expenditure rose by 453 per cent.) Such comparisons suggest either an increasing inability of the parties to raise funds to the maximum or an unwillingness to do so, because of a perception that the effort was not worthwhile. However, differences in the numbers of constituencies, the number of parties contesting each, and the pattern of marginality make such conclusions somewhat tentative. Thus, attention here focuses on contest types (as defined above, p. 53) and on incumbency, following the party fund-raising and candidate fund-raising hypothesis set out in the previous chapter (p. 48).

PARTY INCUMBENCY AND THE PATTERN OF SPENDING

According to the party fund-raising hypothesis, a party's ability to raise funds for a general election campaign is much greater if it holds the seat that if it does not. The validity of this general hypothesis is readily tested by comparing the mean expenditure (as a percentage of the allowed maximum) in seats held by a party with that in seats which it does not hold: one-tailed t tests of differences between means are used to evaluate the statistical significance (at the 0.05 level) of the differences identified. The data used in these tests for the period 1951–79 are set out in Tables 4.2 and 4.4–4.6; those for 1983 are in Table 4.3.

Three-party contests

The general impression given by the means in Table 4.2 is that the amounts spent, as percentages of the maximum, declined considerably, especially during the 1970s. The Conservative party spent

63

on average 94.8 per cent in the seats that it held in 1951, with a standard deviation of only 7.3; 28 years later, the mean had fallen by 10 percentage points and the standard deviation had increased by more than 50 per cent of its 1951 value. The decline in Conservative spending was even greater in the seats held by Labour (from 87.8 to 69.7), with again an increase in its variability, according to the standard deviations. The same general trends occurred for spending by the Labour party and, at a much lower level, for the Liberal party too.

Table 4.2: Summary statistics of spending: Three-party contests (number of constituencies in brackets)

| | | Conservative-held | | | Labour-held | |
		Mean	SD		Mean	SD
1951	(53)			(30)		
Conservative		94.8	7.3		87.8	14.4
Labour		76.1	19.1		86.7	16.1
Liberal		56.1	22.7		49.3	20.0
1959	(67)			(11)		
Conservative		95.8	9.6		77.2	22.2
Labour		78.8	22.0		77.5	20.6
Liberal		63.4	21.7		48.3	19.1
1964	(147)			(26)		
Conservative		95.7	6.4		80.7	23.0
Labour		84.7	19.7		89.7	19.1
Liberal		78.0	20.0		50.8	26.4
1966	(189)			(61)		
Conservative		92.0	7.2		83.7	18.0
Labour		73.2	19.3		84.3	16.4
Liberal		55.7	21.5		43.1	23.0
1970	(160)			(53)		
Conservative		85.6	10.5		83.7	17.7
Labour		58.2	17.5		78.0	11.2
Liberal		44.7	18.3		35.2	21.4
1974	(262)			(185)		
Conservative		90.5	10.2		72.1	22.1
Labour		66.6	20.7		78.7	17.1
Liberal		58.7	23.3		32.9	18.7
1979	(253)			(243)		
Conservative		84.8	11.8		69.7	23.0
Labour		55.8	22.2		74.8	19.2
Liberal		43.6	22.9		21.1	15.1

Differences between the parties have been accentuated over the period. In Conservative-held seats, the incumbent party always outspent its Labour challenger, but whereas the gap between the two was 18.7 percentage points in 1951, in 1979 it was 29 points. Over time, it appears, Labour challengers have been less able and/or willing to raise funds in Tory-held seats. In Labour-held seats, not only has there been little difference between the two parties in their spending (the largest gap was 9 percentage points in 1964), but at two of the elections (1951 and 1970), on average the Conservative challengers outspent the incumbent Labour party. This may indicate a consequence of the unchanging maxima: it clearly implies a greater ability then of the Conservative party to raise campaign funds in seats it did not hold, relative to that of the Labour party. (Interestingly, in both 1951 and 1970, the Conservative party won power from Labour.) Only in 1974 and 1979 did Labour outspend Conservative in those constituencies, and then not by very substantial amounts. In all years except 1951, however, the standard deviation for Labour spending in Labour-held seats was lower than that for Conservative spending therein; the challengers were more variable in their attempts to win through spending (which may reflect on the marginality of seats — see the next chapter).

To test the party fund-raising hypothesis formally, the following are the expectations:

$$CS_C > CS_L$$
$$LS_L > LS_C$$

where CS and LS are Conservative and Labour spending (as percentages of the maxima) respectively, and the subscripts C and L indicate Conservative- and Labour-held seats respectively. The results of the t tests were

	$CS_C > CS_L$	$LS_L > LS_C$
1951	2.45 (s)	2.65 (s)
1959	2.61 (s)	0.18 (ns)
1964	2.07 (s)	1.20 (ns)
1966	3.48 (s)	4.37 (s)
1970	0.73 (ns)	9.51 (s)
1974	10.53 (s)	6.77 (s)
1979	9.12 (s)	10.19 (s)

where (s) indicates significant at the 0.05 level and (ns) indicates not significant. Only in 1970 was the hypothesis invalid for Conservative; for Labour it was invalid in 1959 and 1964. In general, therefore, the hypothesis is sustained, and the ability of parties to raise more funds in the constituencies which they hold than in others is confirmed.

What of the Liberal party, which was the challenger in all of the constituencies? According to the reasoning set out in the party fund-raising hypothesis, there should be no difference in its spending between Conservative-held and Labour-held seats. Formally, the hypothesis (to set against the null hypothesis of no difference) is

$$BS_C \neq BS_L$$

where BS is Liberal spending and the subscripts are as before. Two-tailed t tests are used to evaluate this, with the following results:

1951	+ 1.39 (ns)
1959	+ 2.29 (s)
1964	+ 4.92 (s)
1966	+ 7.91 (s)
1970	+ 2.88 (s)
1974	+12.93 (s)
1979	+12.89 (s)

Thus, in all but the first election studied, Liberal spending was on average significantly greater in Conservative-held than in Labour-held seats, especially at the last two elections, when on average Liberals were spending least. This suggests: (1) that the social composition of Conservative-held seats provides a better foundation for fund-raising by Liberal candidates than does that of Labour-held constituencies (recall that the Liberal party is almost entirely dependent on locally raised campaign funds); and (2) that Liberals perceived greater potential benefits from spending in Conservative- than in Labour-held seats, which reflects the growing electoral polarisation of England and the greater success of the Liberals in the Conservative heartlands of the southeast than in the Labour strongholds of the north (Johnston, 1985a).

At the 1983 election, all of the constituencies had Three-party contests. The general pattern of spending was as hypothesised (Table 4.3). The Conservative party spent on average 25 percentage points more in Conservative-held than in Labour-held seats (producing a t

66

Table 4.3: Summary statistics of spending: 1983 general election (number of constituencies in brackets)

		Conservative		Labour		Alliance	
		Mean	SD	Mean	SD	Mean	SD
Conservative-held seats	(323)	85.1	12.3	58.6	24.8	62.7	23.5
Liberal-contested	(184)	84.2	13.3	52.5	22.7	58.6	23.6
SDP-contested	(139)	86.3	10.9	66.6	25.2	68.2	22.2
Labour-held seats	(187)	60.7	23.0	76.5	19.1	48.1	28.3
Liberal-contested	(76)	59.0	23.7	75.6	20.6	32.5	23.4
SDP-contested	(111)	61.2	22.6	77.1	18.2	58.7	28.8

value of 13.4), and the Labour party spent 18 points more in the constituencies which it was defending (a t value of 9.10). The Alliance candidates spent more challenging in Conservative- than in Labour-held seats (a difference of 14 percentage points; t = 5.97); in both types of seat, the SDP spent significantly more than the Liberal party (t values of 3.73 and 6.81 in Conservative- and Labour-held seats respectively).

For the Conservative party, the Alliance party which it was facing made no significant difference to the amount that it spent, neither in the constituencies which it held, nor in those held by the Labour party. For Labour, the same was true in Labour-held, but not in Conservative-held, seats. In the latter, Labour spent significantly more (t = 5.18) where the other challenger was the SDP than where it was faced by the Liberal party, the implication being that local Labour parties were more determined to stave off the challenge of the SDP (led as it was by Labour defectors), which might relegate their candidates to third place in the Tory strongholds, than they were to counter the Liberal challenge. An alternative interpretation is that in the allocation of constituencies within the Alliance, the SDP got more of the seats in which Labour was expected to do well, and thus where Labour spent above-average amounts. The vote ratios are consistent with this argument: in Labour-held seats, the Labour:Conservative average ratio in 1979 was 1.69 for the seats contested by the Liberal party in 1983, and 1.58 for those contested by the SDP; in Conservative-held seats, the Conservative:Labour average ratios were 2.63 and 1.81 respectively. On average, the SDP was allocated the constituencies where Labour and Conservative were closer, and the changes of a 'third party' correspondingly slimmer (see Curtice and Steed, 1983).

Two-party contests

Two-party contests between Conservative and Labour candidates only (i.e. no Liberal candidates, though there may have been 'other' candidates from minor parties) only occurred in substantial numbers up to 1970 and, as Table 4.4 shows, were more common in Labour- than in Conservative-held seats. Overall, the patterns of spending for the two parties were similar to those in the Three-party contests, with the Conservative party spending more, and having much smaller standard deviations, in the seats that it held than in the Labour-held seats. Labour has outspent Conservative in Labour-held seats at all elections since that of 1951.

Table 4.4: Summary statistics of spending: Two-party contests (number of constituencies in brackets)

		Conservative-held			Labour-held	
		Mean	SD		Mean	SD
1951	(24)			(64)		
Conservative		93.8	9.2		85.9	16.5
Labour		73.8	20.4		81.9	15.5
1959	(136)			(168)		
Conservative		95.5	9.9		82.8	18.6
Labour		81.3	22.2		87.5	18.9
1964	(54)			(116)		
Conservative		95.2	5.5		76.9	17.1
Labour		92.4	19.7		91.7	16.9
1966	(23)			(140)		
Conservative		93.0	6.6		74.8	19.0
Labour		80.0	18.9		83.6	14.6
1970	(22)			(150)		
Conservative		88.9	9.6		68.2	21.5
Labour		61.8	19.2		73.7	18.6
1974 (October)	(0)			(0)		
1979	(0)			(0)		

The formal hypotheses are as for the previous section. The results show the following:

	$CS_C > CS_L$	$LS_L > LS_C$
1951	2.80 (s)	2.73 (s)
1959	7.59 (s)	2.58 (s)
1964	2.39 (s)	−0.22 (ns)

| 1966 | 8.51 (s) | 0.85 (ns) |
| 1970 | 7.56 (s) | 2.67 (s) |

The party fund-raising hypothesis is thus fully confirmed for the Conservative party, but far from so for the Labour party. At two of the elections, there was no significant difference between Labour spending in the two types of seats, and indeed, in 1959, the party raised and spent more in the Conservative-held constituencies. The reasons for this are not entirely clear; they are explored further in a general discussion at the end of the next chapter.

Liberal-exit contests

Contests which involve the non-appearance of a Liberal candidate following the presence of one at the previous election declined substantially in number over the period, and there were none in 1974 and 1979 (Table 4.5). For most purposes these can be equated with the Two-party contests discussed above, since the Liberal party probably did not perform well in the first election of the pair, leaving relatively few votes which either of the other parties could expect to

Table 4.5: Summary statistics of spending: Liberal-exit contests (number of constituencies in brackets)

		Conservative-held			Labour-held	
		Mean	SD		Mean	SD
1951	(168)			(155)		
Conservative		95.1	8.4		92.8	11.0
Labour		72.8	19.0		84.1	16.6
1959	(6)			(8)		
Conservative		97.0	13.1		86.7	11.3
Labour		96.1	13.1		85.4	19.7
1964	(6)			(11)		
Conservative		88.1	11.3		83.8	13.0
Labour		78.2	21.9		91.6	12.9
1966	(35)			(37)		
Conservative		92.8	6.2		83.1	17.0
Labour		80.6	18.1		86.2	15.5
1970	(21)			(35)		
Conservative		85.4	13.5		74.5	23.7
Labour		66.3	17.2		76.6	19.9
1974 (October)	(0)			(0)		
1979	(0)			(0)		

obtain through campaigning. (Recall that, according to some analysts — e.g. Himmelweit *et al.*, 1985 — during this period a Liberal vote was much more likely to be a protest vote against one or both of the other two parties than to reflect a stable commitment to that party.) However, at all five elections the Conservative party did spend more on average in Labour-held seats with Liberal-exit contests than in seats with two-party contests (compare Tables 4.4 and 4.5), suggesting that it perceived greater potential benefits from additional campaigning in the former case, to win over those who voted Liberal previously. (Only two of the differences were statistically significant, however; in 1951 and 1966.) In Conservative-held seats, the Conservative party tended to spend more on Two-party than on Liberal-exit contests, but never significantly so. Labour tended to spend less on Two-party than on Liberal-exit contests in the seats that it held, though only significantly so in 1951 and 1966, suggesting that it too perceived greater electoral benefits from campaigning to win the support of former Liberal than former Conservative voters; in Conservative-held seats, also, it spent more on the Liberal-exit contests in most cases, though only significantly so in 1959.

With regard to the party fund-raising hypothesis (i.e. $CS_C >$ CS_L; $LS_L > LS_C$), the expected relationship was observed for each of the five elections in Conservative-held seats, though it was statistically insignificant in 1964. In Labour-held seats, the expected relationship was present, and statistically significant, in the end-years of the sequence only (1951 and 1970); in 1959, Labour spent more on challenging in Conservative-held seats, on average, than it did defending its hold in Labour-held seats.

Liberal-entry contests

Liberal-entry contests involve the third party providing a candidate in what was previously a two-party contest, thereby putting greater pressure on Conservative and Labour and suggesting that they should respond by spending more on the campaign. In addition to the standard hypotheses, therefore, this further expectation is also investigated.

The three standard hypotheses are:

$$CS_C > CS_L$$
$$LS_L > LS_C$$
$$BS_C \neq BS_L$$

Table 4.6: Summary statistics of spending: Liberal-entry contests (number of constituencies in brackets)

		Conservative-held			Labour-held	
		Mean	SD		Mean	SD
1951	(2)			(1)		
Conservative		97.8	9.2		97.5	–
Labour		77.2	4.3		92.7	–
Liberal		75.4	45.6		67.3	–
1959	(84)			(29)		
Conservative		96.0	8.8		90.8	13.1
Labour		78.5	20.9		89.0	16.5
Liberal		58.0	20.3		56.6	24.6
1964	(109)			(39)		
Conservative		93.8	7.9		79.0	16.6
Labour		87.3	18.5		87.6	18.0
Liberal		61.1	24.0		40.1	15.9
1966	(15)			(9)		
Conservative		92.0	5.1		78.2	20.2
Labour		81.8	18.8		86.8	17.4
Liberal		37.1	16.9		36.0	21.8
1970	(16)			(45)		
Conservative		84.5	10.0		83.2	17.1
Labour		60.8	15.4		80.8	13.1
Liberal		40.2	25.1		26.2	14.2
1974 (October)	(6)			(53)		
Conservative		85.4	8.0		56.3	21.3
Labour		70.9	20.1		72.3	17.7
Liberal		49.8	22.9		22.7	11.6
1979	(0)			(1)		
Conservative		–	–		94.1	–
Labour		–	–		97.5	–
Liberal		–	–		22.6	–

The first of these hypotheses was confirmed as statistically significant in 1959 ($t = 1.96$), 1964 (5.29), 1966 (1.90) and 1974 (6.27), but not 1970 ($t = 0.36$). The second was confirmed for 1959 and 1970 only (the t values for the five elections — Table 4.6 — were 2.71, 0.19, 0.63, 4.50 and 0.15). The third hypothesis was substantiated in 1966 (+6.07), 1970 (+2.05) and 1974 (+2.53), in each case indicating that Liberal candidates spent more, on average, contesting Conservative- than Labour-held seats.

For the Conservative and Labour parties, the entry of a Liberal candidate meant an extra contestant for the available votes. Their response, one would expect, should have been to spend more to counter this new electoral threat, relative to seats where no such additional factor was operating. Comparing Liberal-entry contests to both Three-party and Two-party contests, therefore, the formal hypotheses are:

$$CS_{EN} > CS_2 \qquad CS_{EN} > CS_3$$
$$LS_{EN} > LS_2 \qquad LS_{EN} > LS_3$$

where 2 and 3 refer to Two- and Three-party contests respectively, and EN refers to Liberal-entry contests.

For Conservative-held seats, there is little evidence in support of these expectations. The hypotheses were tested on four elections (1959, 1964, 1966 and 1970), giving t values as follows:

$CS_{EN} > CS_2$	$0.39, \ -1.40, \ -0.51, \ -1.32$
$CS_{EN} > CS_3$	$0.13, \ -2.05, \quad 0.00, \ -0.41$
$LS_{EN} > LS_2$	$-0.94, \ -1.57, \quad 0.28, \ -0.57$
$LS_{EN} > LS_3$	$-0.09, \quad 1.08, \quad 1.65, \quad 0.60$

Only one of the t values has the expected sign and is statistically significant — for 1966 when Labour spent more in Liberal-entry than in Three-party contests. In general, both parties spent more in the relatively certain environments of Two- and Three-party contests than in countering a new Liberal threat, but the differences were rarely statistically significant. In Labour-held seats, the pattern was somewhat different. The t values are:

$LS_{EN} > LS_2$	$0.44, \ -1.24, \quad 0.51, \quad 2.85$
$LS_{EN} > LS_3$	$1.59, \ -0.44, \quad 0.38, \quad 1.11$
$CS_{EN} > CS_2$	$2.79, \quad 0.67, \quad 0.46, \quad 4.80$
$CS_{EN} > CS_3$	$1.83, \ -0.32, \ -0.73, \ -0.14$

The majority of the t-values are positive, indicating differences in the hypothesised direction. Only four of them are statistically significant, however, with three referring to Conservative spending. Thus, although in Labour-held seats the emergence of a Liberal candidate did appear to stimulate extra campaign spending by the other parties, on most occasions the extent of that extra spending was usually slight.

For the Liberal party, the differences should be in the opposite direction, with greater spending in the constituencies with continuity of candidates; new candidates in Liberal-entry contests would either have to rebuild a moribund party organisation or start from scratch, which would make fund-raising more difficult. This suggests that $BS_3 > BS_{EN}$, which is generally confirmed by the t values:

	1959	1964	1966	1970	1974
Conservative-held	1.55	5.95	3.89	0.68	–
Labour-held	−1.09	1.82	0.86	2.46	4.81

In summary

In general, the party fund-raising hypothesis is substantiated by the analyses presented here, though not in every case. Auxiliary hypotheses, referring to different types of contest, have not in general been substantiated, however.

CANDIDATE-INCUMBENCY AND THE PATTERN OF SPENDING

The candidate fund-raising hypothesis suggests that if a party is fielding a candidate who contested the constituency at the previous election, that person should be able to raise more than a new candidate. This applies both to incumbent MPs, those who won at the previous election, and to incumbent-losers, who nevertheless have continued to 'nurse' the constituency during the term of the Parliament. Formally, therefore, this suggests:

(1) in Conservative-held seats

$$CS_I > CS_N$$

(2) in Labour-held seats

$$CS_I > LS_N$$

(3) in Three-party contests

$$BS_I > BS_N$$

where I is an incumbent candidate and N is a non-incumbent candidate.

In addition, it could be suggested that each party might react in its spending to whether the other has an incumbent candidate: for example, in a Conservative-held seat, the Labour party may spend more campaigning against a non-incumbent, especially if the Labour candidate is an incumbent, than against an incumbent, because the chances of victory are perceived to be slightly greater. Because of that, the Conservative party may spend more where it is fielding a new candidate than where it is fielding an incumbent, especially if its new candidate is facing a Labour incumbent. Thus,

(1) in Conservative-held constituencies

$$CS_{CNI, LI} > CS_{CNI, LNI} > CS_{CI, LI} > CS_{CI, LNI}$$
$$LS_{LI, CNI} > LS_{LNI, CNI} > LS_{LI, CI} > LS_{LNI, LI}$$

(2) in Labour-held constituencies

$$LS_{LNI, CI} > LS_{LNI, CNI} > LS_{LI, CI} > LS_{LI, CNI}$$
$$CS_{CI, LNI} > CS_{CNI, LNI} > CS_{CI, LI} > CS_{CNI, LI}$$

where CI, LI are Conservative and Labour incumbent candidates, respectively, and CNI, LNI are Conservative and Labour non-incumbent candidates.

Unfortunately, testing these hypotheses is not always possible. For several of the elections the number of constituencies in certain categories is small, for two reasons: firstly, the shorter the inter-election period, the greater the probability of all incumbents, especially MPs, standing again; and secondly, the relatively small number of losing candidates who stand again in the same constituency if the inter-election period is more than a year. (Many move to a more likely success, after 'proving' themselves in a relatively hopeless seat.)

Incumbent vs non-incumbent: intra-party

For both the Conservative and the Labour parties, the hypothesis that incumbent candidates raise more than non-incumbent candidates is resoundingly rejected. For the Conservative party in Conservative-held seats, 16 tests were carried out. *None* produced a statistically significant difference in the expected direction, and three — for Three-party contests in 1964 and 1966 and Two-party

Table 4.7: Spending by each party in 1983 according to party and candidate incumbency and Alliance party

Winning party	Conservative Mean	SD	Labour Mean	SD	Alliance Mean	SD
Conservative						
All Seats	85.1	12.3	58.6	24.8	62.7	23.5
Liberal (N = 184)	84.2	13.3	52.5	22.7	58.6	23.6
SDP (N = 139)	86.3	10.9	66.6	25.2	68.2	22.2
Conservative Inc.	84.3	12.7	57.0	24.5	62.4	23.0
Liberal (N = 150)	83.4	13.7	50.8	21.9	58.3	23.4
SDP (N = 110)	85.7	11.3	64.6	25.6	68.1	21.2
No Conservative Inc.	88.1	9.9	66.5	24.5	64.1	25.7
Liberal (M = 34)	87.7	10.6	60.1	24.7	60.3	24.9
SDP (N = 29)	88.5	9.1	74.1	22.3	68.6	26.3
Labour						
All seats	60.7	23.0	76.5	19.1	48.1	28.3
Liberal (N = 76)	59.9	23.7	75.6	20.6	32.5	23.4
SDP (N = 111)	61.2	22.6	77.1	18.2	58.7	28.8
Labour Inc.	60.2	22.7	75.9	19.4	43.7	25.4
Liberal (N = 57)	58.7	22.2	74.6	21.8	32.8	22.2
SDP (N = 65)	61.5	23.2	77.2	17.2	53.3	24.2
No Labour Inc.	61.6	23.7	77.5	18.7	56.2	32.2
Liberal (N = 19)	63.5	27.9	78.6	16.6	31.7	27.4
SDP (N = 46)	60.8	22.0	77.1	19.7	66.3	28.5

contests in 1959 — produced significant t values in the opposite direction to that hypothesised. Fourteen tests were carried out for Labour spending in Labour-held seats: one produced a significant t value in the expected direction (Two-party contests in 1966), and two produced comparable but negative t values (Two-party contests in 1964; Liberal-exit contests in 1951).

In the 1983 analyses, candidate-incumbency refers only to the party holding each seat, so no analyses can be undertaken of the losing incumbents who nursed a constituency. (This was because of the boundary changes. MPs were classed as incumbents if they had previously represented at least part of the constituency for which they stood; no attempt was made to characterise non-MPs similarly.) For the Conservative party there were no significant differences in spending in the Conservative-held seats (Table 4.7), whether or not it was fielding an incumbent and whether it was facing a Liberal or an SDP candidate for the Alliance. Labour candidates spent significantly more challenging a non-incumbent than an incumbent, however, suggesting that their campaign effort was increased where they perceived a greater potential for success. The same was not true for Conservative spending in Labour-held seats, however.

Within the Alliance, Liberal candidates spent approximately the same amount whether facing an incumbent or a non-incumbent, though there was a significant difference between Conservative- and Labour-held seats in their average spending. The SDP did not differentiate between Conservative incumbents and non-incumbents either, but in Labour-held seats it spent on average 13 percentage points more where Labour was not fielding an incumbent than where it was so doing. Again, this suggests a greater propensity to spend in the equivalent of the 'open seat' in the USA (p. 31). However, in a number of Labour-held seats without an incumbent, in fact the MP elected at the last election was now the SDP candidate, who defected to that party after its formation in 1981. These SDP incumbents were provided with as much campaign money as they needed by the party headquarters (David Owen spent 93 per cent in Plymouth Devonport, Bill Rodgers spent 99 per cent in Stockton North, as did Ian Wrigglesworth in Stockton South, and John

Table 4.8: Summary statistics of spending by Liberal party incumbent and non-incumbent candidates in Three-party contests (number of constituencies in brackets)

		Conservative-held			Labour-held	
		Mean	SD		Mean	SD
1951						
Incumbent	(23)	54.0	22.6	(6)	48.5	16.7
Non-incumbent	(14)	59.6	23.3	(8)	50.3	25.5
1959						
Incumbent	(5)	68.5	21.3	(0)	–	–
Non-incumbent	(62)	63.0	21.9	(11)	48.3	19.1
1962						
Incumbent	(41)	78.0	23.1	(4)	53.3	39.7
Non-incumbent	(106)	78.0	18.3	(22)	50.3	24.5
1966						
Incumbent	(82)	56.1	22.3	(24)	34.1	22.4
Non-incumbent	(107)	55.4	20.9	(37)	48.9	21.7
1970						
Incumbent	(26)	50.6	15.7	(11)	42.2	26.4
Non-incumbent	(134)	43.5	18.6	(42)	33.3	19.7
1974 (October)						
Incumbent	(186)	60.6	23.3	(115)	34.7	20.1
Non-incumbent	(76)	53.9	22.8	(70)	29.8	15.7
1979						
Incumbent	(52)	51.7	26.9	(39)	25.7	19.0
Non-incumbent	(200)	41.5	21.3	(200)	20.2	14.2

Table 4.9: Summary statistics of spending by Liberal party in Liberal-entry contests (number of constituencies in brackets)

| | | Conservative-held | | | Labour-held | |
		Mean	SD		Mean	SD
1951	(2)	75.4	45.8	(1)	67.3	–
1959	(84)	58.0	20.3	(29)	56.6	24.6
1964	(109)	61.1	24.0	(39)	40.1	15.9
1966	(15)	37.1	16.9	(9)	36.0	21.8
1970	(16)	40.2	25.1	(45)	26.2	14.2
1974	(6)	49.8	22.9	(53)	22.7	11.6

Cartwright spent 98 per cent in Woolwich.) They will have substantially raised the average SDP expenditure in Labour-held seats without a Labour incumbent, therefore.

For the Liberal party prior to 1983, an incumbent candidate who 'nursed' a constituency between elections would probably have been more valuable than either a comparable Conservative or Labour losing-incumbent in maintaining party vitality, and this should have been reflected in campaign activity; Liberal incumbents should have been able to raise substantially more than new candidates. The data in Table 4.8 show that this was not so in Three-party contests in the early years of the period studied, but that it was so from 1970 onwards — the period of the Liberal revival when, presumably, incumbents were able to mobilise support because of the (less faint?) chances of electoral success. Thus, there are significant t values for Conservative-held seats from 1970 onwards, but not before, and for Labour-held seats in 1974 and 1979. (Note that in 1966, the non-incumbent Liberals in Labour-held seats substantially outspent the incumbents, however.) In every year, spending was greater by each type of Liberal candidate (incumbent or non-incumbent) in Conservative-held than in Labour-held seats — with nine of the twelve comparisons possible producing statistically significant t values (these also incorporate the Liberal-entry contests; Table 4.9).

It seems that non-incumbent Liberal candidates contesting a constituency for the first time face difficulties raising campaign funds. However, most candidates will inherit an active party organisation for that task. Those contesting Liberal-entry seats (all of whom are by definition non-incumbents) may lack even that foundation, however, making for even greater difficulties in raising funds locally. (Recall that local fund-raising is the predominant

77

source for Liberal candidates.) This suggests that spending by non-incumbent Liberals in Three-party contests (Table 4.8) should be greater than that of Liberals in Liberal-entry contests (Table 4.9). In general this has been so, as the following t values indicate:

	1959	1964	1966	1970	1974
Conservative-held	1.40	5.79	3.71	0.49	0.39
Labour-held	−1.09	1.72	1.52	1.89	2.86

In Conservative-held constituencies, however, there was much less difficulty in the 1970s, suggesting either that a party organisation was available, or that the chances of success were perceived as such as to stimulate substantial fund-raising efforts in the Liberal-entry contests. In all five elections, Liberal entrants spent more in Conservative- than in Labour-held seats, significantly so in 1964, 1974 and 1979 (with the last two further supporting the conclusion of the previous sentence.)

Incumbent vs non-incumbent: inter-party

The hypotheses for this section suggest sequences of spending levels, with three significant differences between each adjacent pair. Testing them was not feasible in many cases, however, because of the small numbers of observations. The reasons for this are straightforward. When a pair of elections is relatively close together (i.e. 1950–1951; 1964–1966; 1974 (February)–1974 (October)), very few incumbent MPs do not stand at the second election, whilst very many of the defeated candidates stand again; thus, there are few non-incumbents for either party. When there are several years between a pair of elections, on the other hand, relatively few of the defeated candidates stand again, so that almost all of the challengers are non-incumbents. Thus, for example, in the Three-party contests, the full sequence was present (i.e. with at least five entries only in each cell) for none of the elections in Conservative-held seats and for two in the Labour-held seats. A further four full sequences only were obtained — three of them relating to the 1959 election. Thus, most of the tests referred to parts of the sequence only.

The data on which the t tests were based are not presented here, since the great majority of the tests indicated no significant differences. In the Conservative-held seats with Three-party contests, for example, 22 tests were conducted. For Conservative spending, only one t value was significant with the correct sign, and for Labour the number was only two. (Interestingly, for both parties spending was greater in 1979 in contests between two incumbent candidates than in constituencies where the incumbent party fielded an incumbent candidate but the challenger's candidate was new to the constituency.) Twenty-two tests were also performed on Three-party contests in Labour-held seats. Only two produced the expected significant t value; both Labour and Conservative spent more in 1979 where a new Labour candidate was facing a Conservative incumbent than in seats where both parties fielded a non-incumbent candidate.

Of the other tests conducted, none of the 22 in Conservative-held seats with contests other than Three-party was significant. A further 44 tests were possible in Labour-held seats, with six only resulting in statistically significant t values.

CONCLUSION

This chapter presents only the first part of the evaluation of the models developed in Chapter 3; four more hypotheses are tested in the next chapter. Hence, detailed interpretation is left to a later stage. For the present, however, it can readily be concluded that the party fund-raising hypothesis has a great deal of support but the candidate fund-raising hypothesis has virtually none. Whether a party holds a seat or not has generally been crucial to how much it raised for the campaign there; who its candidate may be was irrelevant.

5

Electoral Context, Spending, and Electoral Outcome

The previous chapter looked at the pattern of spending at the eight general elections studied, and tested two of the hypotheses set out in Chapter 3. The present chapter extends the analysis, by investigating the full models enunciated in Figures 3.2–3.5. This involves testing four further hypotheses. The continuity hypothesis (which is discussed in a separate section) states simply that a party's share of the vote — as indexed by the vote ratios — at one election will be closely related to its share at the previous contest. The marginality hypothesis suggests that the closer the vote ratio at the first election, the more a party will spend at the second, because of perceived higher benefit:cost ratios. The party-spending hypothesis claims that the more that a party spends, *ceteris paribus*, the greater its share of the vote, and the candidate incumbency hypothesis suggests that incumbents will increase their share of the vote, relative to non-incumbents.

The style of reporting adopted here does not focus on the particular hypotheses in turn. Rather, the focus is on the various models, which refer to the four different types of contest: Three-party (Conservative, Labour, Liberal); Two-party (Conservative, Labour); Three-party with Liberal Entry (i.e. a Liberal candidate contested the second of the elections but not the first); and Two-party with Liberal exit (a Liberal candidate contested the first of the elections in the pair, but not the second). A section is assigned to each, and within each there is a separate analysis of Conservative-held and Labour-held seats. As Table 5.1 shows, not every contest type was present in sufficient numbers at every election for an analysis to be undertaken, however; if the contest type was represented in less than 20 constituencies, it was not included in the analyses.

Table 5.1: The number of constituencies in each contest type at each election

	Three-party	Two-party	Liberal-exit	Liberal-entry
1951				
Conservative-held	53	24	168	2
Labour-held	30	64	155	1
1959				
Conservative-held	67	136	6	84
Labour-held	11	168	8	29
1964				
Conservative-held	147	54	6	109
Labour-held	26	116	11	39
1966				
Conservative-held	189	23	35	15
Labour-held	61	140	37	9
1970				
Conservative-held	160	22	21	16
Labour-held	53	150	35	45
1974 (October)				
Conservative-held	262	0	0	6
Labour-held	185	0	0	53
1979				
Conservative-held	253	0	0	0
Labour-held	243	0	0	0
1983				
Conservative-held	323	0	0	0
Labour-held	187	0	0	0

The results of the analyses are presented in three ways. The models (Figures 3.3–3.5) require two sets of regression analyses: the first set regresses the pattern of spending by each of the parties on marginality (the vote ratio between the two parties at the first election) and whether that party's candidate was an incumbent (i.e. had contested the constituency at the previous election); the second regresses the pattern of votes at the second election (the ratio between the votes won by the two parties) on the pattern at the first, on the level of spending by each party (as a percentage of the allowed maximum in the constituency), and on the candidate-incumbency variables. Thus, in the contests involving three parties (Three-party contests and Liberal-entry contests), there are six regressions — the dependent variables are the level of spending by each party and the vote ratio between each pair of parties; in the contests involving two parties (Two-party contests and Liberal-exit contests), there are only three. The full regression equations are given in a series of tables (Tables 5.1A–5.31A) which are appended to this chapter. These tables give each of the terms of the equation (the a and partial b coefficients together with, for the latter, the t-values and partial beta coefficients) plus a measure of the goodness-of-fit (R^2, adjusted for degrees of freedom). (Note that in the regressions on the vote ratios, for those contest types involving three parties, the impact of spending by each is assessed, although one is not included in the model — for instance, the impact of Liberal spending on the Conservative:Labour ratio. This impact is outside the model, since whether it exists, and what direction it would take, could not be specified *a priori*. Its inclusion is thus exploratory, enquiring as to whether the amount of campaign effort by a third party did have any impact on the vote ratio for the other two, across the system as a whole.)

The second form of presenting the results comprises a series of diagrams (Figures 5.1–5.17) which indicate the statistically significant elements of the models in each case. Since in all of the models there was a directional element to the relationships, significance was assessed by one-tail t tests on the unstandardised regression co-efficients. (A significance level of 0.05 was used.) The role of significance testing in such analyses is somewhat controversial (see Hay, 1985). Although the models clearly call for confirmatory rather than exploratory data analyses (Johnston and Wrigley, 1987), since they involve clearly-specified hypotheses, they are not being tested on samples; the full set of observations (i.e. the population) is being analysed in each case. To some, this issue is circumvented by arguing that the populations studied are themselves samples of some larger populations (all possible electors, perhaps). This is not presented here, however, both because it is tenuous and also because no argument is being presented that the models are general and contain putative social-scientific laws. The enquiry is only into whether spending at particular elections was related to the pattern of votes, not into whether spending at elections is always related to the pattern of votes (i.e. this is a piece of empirical science, but not of positivist science — Johnston, 1986d). Statistical significance tests are being employed only to enquire as to whether the postulated relationships are present in a strong form (i.e. the t-values measure the closeness of the scatter of data points around the partial regression lines). The stronger the relationship in this context, the greater the confidence that the hypothesised links are present in the data.

The final format for presenting the results is a series of summary tables included within the text showing the main relationships and indicating the major conclusions.

THE CONTINUITY HYPOTHESIS

Most of the tests in this chapter relate to the pattern of spending — its links to candidate incumbency and to marginality and its impact on the election result. All of these refer to what are expected to be relatively small changes in the pattern of voting from one election to the next. As discussed in Chapters 1 and 3, the electoral geography of England has been extremely stable for over 60 years, including the period of the present study. (For a discussion of that stability and its causes, see Johnston, 1987a.) Thus, it was expected not only that this continuity would be observed, but that it would be the dominant feature of all the regression equations.

To assess the relative importance of this feature of the electoral geography of England, thereby evaluating the continuity hypothesis, the beta coefficients (i.e. the standardised partial regression co-efficients) from the regression equations are employed. These indicate the rate of change in the dependent variable relative to that in the independent variable, when each is standardised to a zero mean and unit variance (Johnston and Wrigley, 1987). Thus, for any multiple regression equation, the relative size of the beta weights indicates the relative importance of the independent variables as influences on the dependent. According to the continuity hypothesis, the beta weights for the variables representing the result of the previous election should be by far the largest.

In the analyses of Conservative-held seats, nineteen separate regressions were computed in which the dependent variable was the ratio between two parties' share of the vote at the second election and one of the independents was the similar ratio for the first election. (Nine of these referred to Three-party contests, with separate analyses for Liberal- and SDP-contested seats in 1983; five referred to Two-party contests; three to Liberal-exit contests; and two to Liberal-entry contests.) The average beta coefficient for the Conservative:Labour ratios was 0.90, with a standard deviation of 0.08. The average beta coefficient for the second-most influential independent variable was 0.13, with a standard deviation of 0.08. Thus, over all 19 separate contests (elections by types), the beta coefficient for continuity was on average seven times larger than that for any other variable — clear evidence of the importance of continuity in the electoral geography of Conservative-held seats.

A very similar result holds for the Labour-held seats, for which there were 20 separate analyses (one less than in Conservative-held seats for Three-party contests but two more for Liberal-entry).

83

Across all twenty the average beta coefficient for the influence of the Labour:Conservative ratio at the first election on the same ratio at the second was 0.93, with a standard deviation of 0.09, whereas for the second-most influential independent variable, the respective values were 0.09 and 0.06. On average, the beta coefficient for continuity was 14 times larger than that for any other variable.

Ratios between the share of the votes won by either Conservative or Labour and that of Liberal could be computed for both elections in a pair for the Three-party contests only, of which there were nine for Conservative-held seats and eight for Labour-held seats. The average beta coefficients for the vote variables and the largest for any independent variable other than the vote ratio (in five cases another variable had a higher beta than did the ratio) were

Conservative-held seats
    Conservative:Liberal ratio 0.55   Other variable   0.32
    Labour:Liberal ratio   0.61   Other variable   0.35
Labour-held seats
    Labour:Liberal ratio   0.53   Other variable   0.27
    Conservative:Liberal ratio 0.47   Other variable   0.34

On average, then, in the contests involving the Liberal party the beta coefficient for the vote ratio was between 1.5 and 1.9 times greater than that for any other independent variable. Continuity was much less predominant as an influence on the result of the second election in these cases, therefore, and other variables (relating to spending in nearly every other case) also had considerable relative impact on the outcome.

The continuity hypothesis is clearly substantiated by these analyses, therefore, especially with regard to the relative share of the votes won by Conservative and Labour. Nevertheless, it is not the sole influence on the election results, particularly with regard to the relative share of the vote won by the Liberal party. The remainder of this chapter focuses on those other influences, with particular reference to the level and impact of campaign spending.

THE THREE-PARTY CONTESTS

The Three-party contests are those in which the Conservative, Labour and Liberal parties each put up a candidate at both of the elections in the pair investigated. Consequently, the only contextual

Figure 5.1: Significant links in the models for Conservative:Labour ratios in Conservative-held seats, Three-party contests

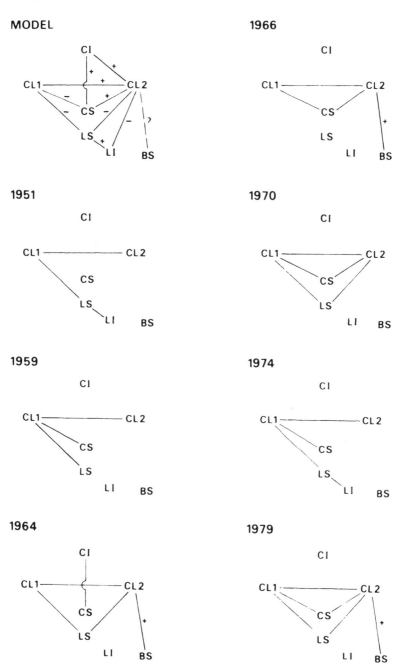

Figure 5.2: Significant links in the models for Labour:Conservative ratios in Labour-held seats, Three-party contests

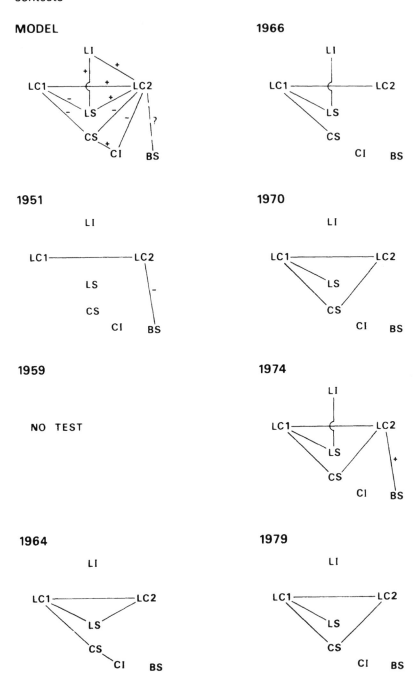

change in the constituency in most cases will relate to the individual candidates, who may or may not be incumbents (unless there is a major political event or controversy in particular constituencies). A stable pattern following the national trend should be apparent, with the level of spending causing some small deviation from that general trend; incumbent candidates might also produce slight variations, obtaining slightly larger percentages of the vote than non-incumbents.

Looking first at the Conservative:Labour components of these contests, Figures 5.1 and 5.2 show the significant links. One clear conclusion stands out; not only did having an incumbent candidate have no significant impact in most cases on the level of spending once marginality (the vote ratio at the first contest) was held constant (only six significant coefficients), but it had no impact at all on the result of the second election (no significant regression coefficients). Extending the results of Chapter 4, therefore, it is clear that the hypotheses relating to candidate incumbency were of virtually no relevance in the elections studied.

The pattern of spending was significantly related to the marginality of the contest in almost every case, especially for the Labour party. In the Conservative-held seats, there was a significant negative relationship at every one of the seven elections between the Conservative:Labour ratio at the first election of the pair and the level of Labour spending; the greater the distance between the two parties (i.e. the lower the marginality), the less that Labour spent. In two cases — the 1951 and 1964 elections — however, the level of Conservative spending was not significantly related to marginality. In the Labour-held seats, all of the expected relationships between marginality and spending were present, for both the challenger party (Conservative) and the party holding the seat (Labour), except at the 1951 election.

Further information on the nature of these significant relationships is provided in Table 5.2. Comparison of the size of the regression coefficients shows that in every case the slope of the relationship was steeper for the challenger party than it was for the incumbent party (a not unexpected finding in many ways, given the relative size of the standard deviations reported in Chapter 4). This difference is especially marked in the Conservative-held seats, where the ratio between the two slope coefficients varied between 3.2 (1979) and 6.9 (1966): in the Labour-held seats, the ratio was never as large as 2.0 and in 1979 was only 1.1. Clearly, Labour challengers have paid much more attention to the winnability of the

Table 5.2: Summary table of regression results — Three-party contests

	1951	1959	1964	1966	1970	1974	1979
Conservative-held seats							
Regression Coefficients (and R^2) for Spending on Marginality							
CS	x(14)	−0.033(8)	x(3)	−0.016(3)	−0.034(3)	−0.017(5)	−0.021(2)
LS	−0.241(45)	−0.158(44)	−0.134(43)	−0.110(39)	−0.117(39)	−0.093(44)	−0.067(29)
BS	−0.114(45)	−0.116(49)	−0.059(15)	−0.077(20)	−0.109(25)	−0.166(30)	−0.214(20)
Regression Coefficients for CL Ratio on Spending							
CS	x	x	x	0.385	0.325	x	0.519
LS	x	x	−0.387	x	−0.341	−0.289	−0.280
Regression Coefficients for CB Ratio on Spending							
CS	x	x	x	x	x	x	x
BS	−3.367	−1.259	−1.515	−0.843	−1.920	−0.813	−1.582
Labour-held seats							
Regression Coefficients (and R^2) for Spending on Marginality							
LS	x(29)		−0.088(14)	−0.078(33)	−0.103(36)	−0.100(20)	−0.084(23)
CS	x(44)		−0.136(36)	−0.131(49)	−0.186(55)	−0.191(48)	−0.096(47)
BS	x(21)		−0.488(37)	x(13)	−0.443(26)	−0.300(12)	−0.306(11)
Regression Coefficients for LC Ratio on Spending							
LS	x		x	x	x	x	x
CS	x		−0.148	x	−0.443	−0.300	−0.306
Regression Coefficients for LB Ratio on Spending							
LS	x		x	x	2.423	x	x
BS	x		x	−2.596	−4.131	−1.928	−4.336

x statistically insignificant

seats, and have raised much more in the most marginal than has been the case with the Conservative party, which has varied much less in what it spends from constituency to constituency (see also Houghton Report, 1976).

These differences are further encapsulated in the relative sizes of the R^2 values, also given in Table 5.2. In all cases, the R^2 value is much larger for the challenger party than for the incumbent party, indicating a greater degree of regularity of concentration of spending in the more marginal seats by the former than by the latter. This is especially the case in seats where the Conservative party was the incumbent, with an average R^2 value of only 0.07 for Conservative spending compared with 0.42 for the Labour challengers. The difference between the two in seats which Labour held was smaller (0.26 for Labour, 0.47 for Conservative), but the same general trend held. In general, incumbent parties were less likely to raise more in the seats that they held than were their challengers, whatever the level of marginality.

Whereas the hypotheses represented by the links for the pattern of spending in the left-hand side of the models in Figures 5.1 and 5.2 were generally upheld, the same was not true of the right-hand side. Spending was related to the pattern of votes at the first election, but the pattern of votes at the second election was less likely to be related to the relative levels of spending. This was especially so with regard to spending by incumbent parties. In the Conservative-held seats, that party's expenditure was significantly related to the vote ratio at three of the seven elections only (1966, 1970 and 1979 — all late in the sequence), whereas in the Labour-held seats, the incumbent party spending was not significantly related to the electoral outcome in any of the six contests analysed. For challengers, spending was apparently a little more effective, being significantly related to the vote ratio in eight cases in all. (Again, it is noteworthy that none of these significant relationships relate to the 1956 and 1959 elections.)

The conclusion to be drawn from this part of the findings, therefore, is that spending by the challenger party was much more likely to influence the electoral outcome than was spending by the incumbent party, especially in the later years of the period studied. The first part of this conclusion is not very surprising: given the higher levels and smaller variation in incumbent than challenger party spending, the lesser impact of the former is a natural consequence. With regard to the second part, the most likely reason for the greater impact at later elections is the greater variation in spending at those elections, especially in the 1970s by the Conservative

Figure 5.3: Significant links in the models for Conservative:Liberal ratios in Conservative-held seats, Three-party contests

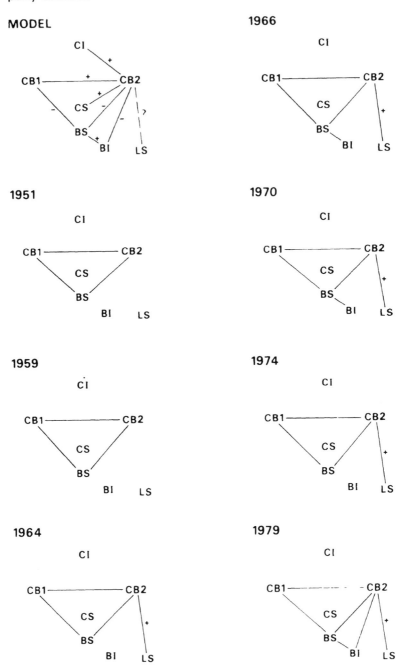

Figure 5.4: Significant links in the models for Labour:Liberal ratios in Conservative-held seats, Three-party contests

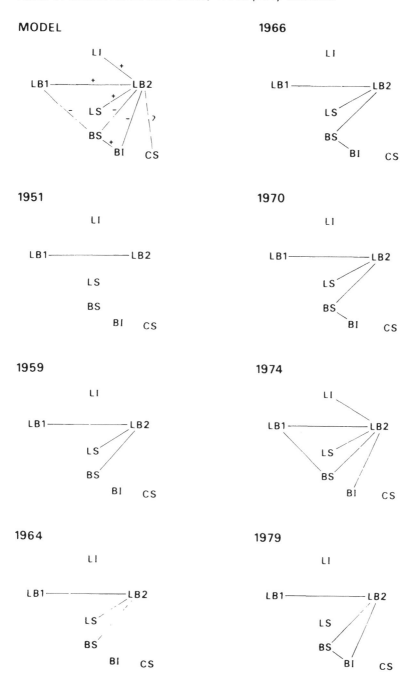

Figure 5.5: Significant links in the models for Labour:Liberal ratios in Labour-held seats, Three-party contests

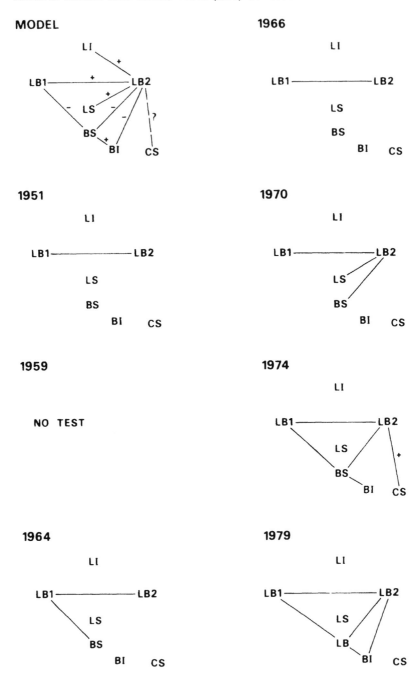

Figure 5.6: Significant links in the models for Conservative: Liberal ratios in Labour-held seats, Three-party contests

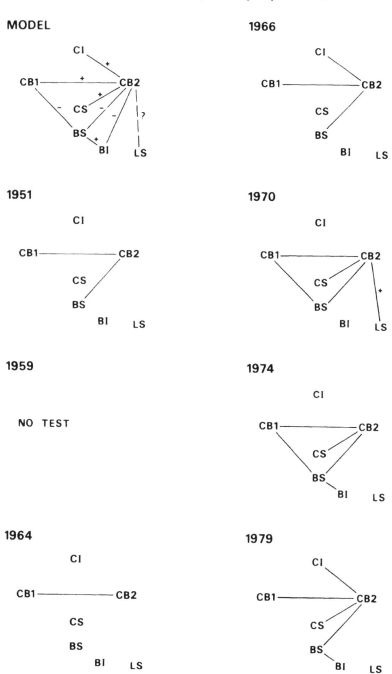

party (Table 4.2) as a consequence of the increases in the allowed maxima.

Turning now to the spending and electoral fortunes of the Liberal party in these contests, Table 5.2 and Figures 5.3–5.6 show results similar to those outlined above for the Conservative:Labour contests. Thus in the Conservative-held seats, the level of Liberal spending was significantly related to the marginality of the Conservative:Liberal contest at each of the seven elections, and the amount that it spent significantly affected the outcome at each. Liberal candidates spent more challenging Conservatives, the smaller the margin between the two parties at the first election of a pair, and were rewarded by a greater proportion of the votes gained by the two parties at the second election; the amount spent by the Conservative party had no counteracting influence, however. In Labour-held seats, on the other hand, the relationships were more equivocal: in only three of the six contests did the Liberal candidates spend more, the more marginal the seat, though in four cases this had a significant influence on the ratio of votes between the two parties (again, in the 1960s and 1970s). Labour spending influenced the Labour:Liberal ratio in only one contest — 1970.

Again, these results are not entirely surprising. One could anticipate that a 'third party' would be more likely to raise money in the more marginal constituencies. Moreover, since it was very much the third party in nearly all of the constituencies (certainly prior to the 1970s), the information which it produced through its campaign would be more likely to have an impact than that of the two 'main parties', which were competing to form the government. In general terms, too, the Liberal party was seen during much of the period as an alternative to the Conservatives, so that its impact on the Conservative:Liberal ratio throughout is understandable; it was only at the later elections, when it was developing its push towards equality with Labour (almost achieved in terms of votes in the 1983 election, in partnership with the SDP) and developing policies that could be perceived as viable alternatives to Labour's (on the welfare state and nuclear disarmament, for example), that it could successfully promote its candidates against Labour incumbents.

One aspect of the contests not covered in the summary tables and figures, but included in the full regression equations, is the impact of spending by the third party (whichever it was) on the vote ratio of the other two — for example, of Liberal spending on the Conservative:Labour ratio. In the Labour-held seats, there is no consistent pattern whatsoever, and only four of the 18 coefficients are statistically significant. In the Conservative-held seats, on the other hand (as shown in the tables appended to this chapter), there are two very clear patterns. The first is that in each of the seven elections, the amount spent by Labour had a positive impact on the Conservative:Liberal ratio — i.e. Labour spending benefited the Conservative party. The reason for this could be that the greater the amount of campaigning by Labour, the less chance the Liberal party had of winning over voters from Labour, and so the better the performance of Conservative relative to Liberal; the latter party did best where Labour's campaign was weakest. Secondly, and consistent with this, in four of the elections (again excluding those in the 1950s), the level of Liberal spending had a significant, positive impact on the Conservative:Labour ratio; the more that the Liberals spent, the better the Conservatives did. In part this may be a statistical artifact, however, since Liberal spending was highest where its ratio with Conservative was lowest and Labour's more likely to be relatively high. Where the Conservative lead over Labour was great, Liberals were able to erode the relative vote-winning capacity of the latter over the former, when they spent a lot on the campaign.

THE TWO-PARTY CONTESTS

The Two-party type was only present in sufficient numbers for regression analysis of five of the seven elections (Table 5.1), and was much more characteristic of the Labour-held than the Conservative-held seats. This supports the suggestion made above that over most of the period studied, the Liberal party was much more of a competitor for votes with the Conservative than with the Labour party. In general, also, the Two-party contests in Labour-held seats were very unequal, with many of them having very large Labour majorities over Conservatives (see the discussion below of Table 5.6), and where the Liberal chances of substantial electoral gains were therefore slight. Perhaps more than the Three-party type, Two-party contests probably contain constituencies where spending is less likely to have an impact and produce deviations from long-term continuity.

With regard to the expected relationships between marginality and spending (the left-hand side of the model in Figures 5.7 and 5.8), all but one are present (the exception being Conservative spending in Conservative-held seats in 1951). As in the Three-party contests, the regression and correlation coefficients were always smaller for the incumbent party than for the challenger (Table 5.3),

Figure 5.7: Significant links in the models for
Conservative:Labour ratios in Conservative-held seats, Two-party
contests

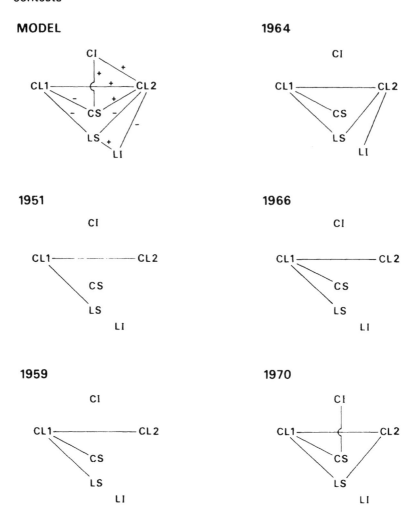

and the ratio between the regression coefficients was much greater
in the Conservative-held than in the Labour-held seats.

The impact of spending, even by the challengers, was much less
substantial in the Two-party than in the Three-party contests,
however, with only five significant regression coefficients out of a
possible total of 20. Three refer to the 1959 election, and four of the
five relate to the challenger party's spending. (The exception was
Labour spending in Labour-held seats in 1959.) Thus, apart from
evidence that the amount spent by Labour significantly reduced the

Figure 5.8: Significant links in the models for
Labour:Conservative ratios in Labour-held seats, Two-party
contests

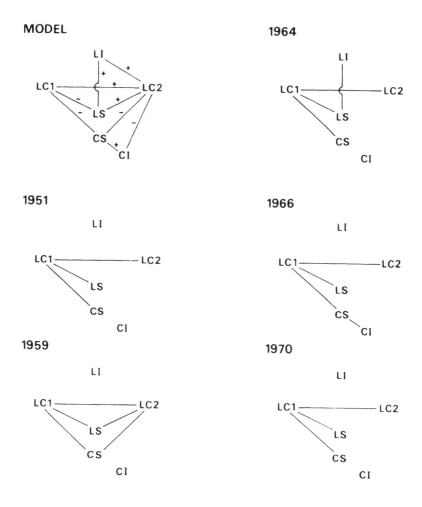

Conservative:Labour ratio in Conservative-held seats in three of the
five elections, in general it must be concluded that spending was of
little import in these contests — despite the clear links between
marginality and amount spent. This suggests that in such contests the
two parties were very largely being judged in the same way in all,
with the volume of local information having little influence on the
voters' evaluations. It is important to note, however, that no Two-
party contests were analysed after 1970, which was the period when
the maxima were raised and spending variability increased.

Table 5.3: Summary table of regression results — Two-party contests

	1951	1959	1964	1966	1970
Conservative-held seats					
Regression Coefficients (and R^2) for Spending on Marginality					
CS	x(14)	−0.104(23)	−0.052(11)	−0.110(15)	−0.219(23)
LS	−0.447(59)	−0.362(58)	−0.338(57)	−0.604(67)	−0.578(51)
Regression Coefficients for CL Ratio on Spending					
CS	x	x	x	x	x
LS	x	−0.328	−0.231	x	−0.843
Labour-held seats					
Regression Coefficients (and R^2) for Spending on Marginality					
LS	−0.065(18)	−0.095(11)	−0.062(7)	−0.034(17)	−0.073(22)
CS	−0.145(62)	−0.169(38)	−0.143(34)	−0.140(42)	−0.163(59)
Regression Coefficients for LC Ratio on Spending					
LS	x	0.272	x	x	x
CS	x	−0.281	x	x	x

x statistically insignificant

THE LIBERAL-EXIT CONTESTS

Liberal-exit contests were present in sufficient numbers for regression analysis at only three of the seven elections studied (1951, 1966 and 1970). In most respects they should be typical of the Two-party contests, except that the Liberal votes cast at the first election of the pair are available, so that spending might be to one or the other party's relative advantage. The results (Figures 5.9 and 5.10; Tables 5.19A–5.21A: summarised in Table 5.4) show this to be the case.

The regression analyses summarised in Table 5.4 are very similar to those in Table 5.3. With one exception (the Conservative party in 1970) each party spent more, the more marginal the seat, and, with one exception also (the size of the correlation and regression coefficients in Labour-held seats in 1951), the general pattern of coefficients was as in the Three- and Two-party contests. However, the impact of spending was even less than in the Two-party contests, with only one significant regression coefficient (for Labour spending in Conservative-held seats in 1970). There is no convincing evidence, therefore, that the two main parties were able to capitalise on the withdrawal of Liberal candidates through their campaign

Figure 5.9: Significant links in the models for Conservative:Labour ratios in Conservative-held seats, Liberal-exit contests

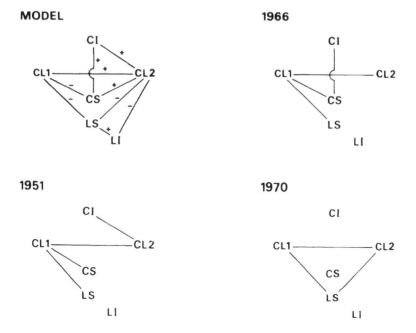

Figure 5.10: Significant links in the models for Labour:Conservative ratios in Labour-held seats, Liberal-exit contests

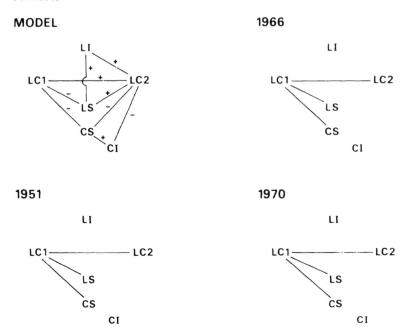

MODEL   1966

1951   1970

Table 5.4: Summary table of regression results — Liberal-exit contests

	1951	1966	1970
Conservative-held seats			
Regression Coefficients (and R^2) for Spending on Marginality			
CS	−0.056(10)	−0.072(20)	x(1)
LS	−0.213(29)	−0.325(37)	−0.173(24)
Regression Coefficients for CL Ratio on Spending			
CS	x	x	x
LS	x	x	−0.397
Labour-held seats			
Regression Coefficients and (R^2) for Spending on Marginality			
LS	−0.111(27)	−0.133(25)	−0.094(21)
CS	−0.067(21)	−0.196(57)	−0.146(38)
Regression Coefficients for LC Ratio on Spending			
LS	x	x	x
CS	x	x	x

x statistically insignificant

strategies. (Of course, the seats in this type were largely those in which the Liberals performed badly in the first contest, so there were few votes to win. Spending did not help. Again, there were no such contests after 1970.)

THE LIBERAL-ENTRY CONTESTS

Liberal-entry contests were present in sufficient numbers for analysis at only two of the elections in Conservative-held seats (1959 and 1964), but at two others (1970 and 1974) in Labour-held seats, reflecting the larger number of Two-party contests in the latter at the earlier elections. The models are similar to those for Three-party contests (Figures 5.11–5.14), except with regard to (1) the prediction of the level of Liberal spending, and (2) the prediction of the vote ratios involving the Liberal party at the second election from the result of the first. For both (Figure 3.5), the ratio between the votes for the other two parties was used as the independent variable. Thus, for example, in Conservative-held seats it was suggested that: (1) the greater the Conservative:Labour ratio at the first election, the

Figure 5.11: Significant links in the models for Conservative:Labour ratios in Conservative-held seats, Liberal-entry contests

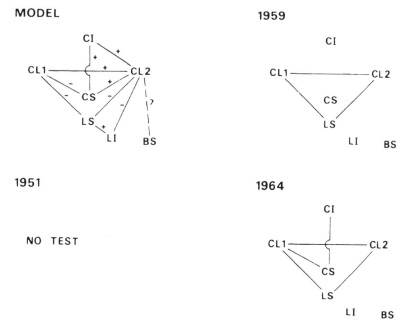

Figure 5.12: Significant links in the models for ratios involving the Liberal party in Conservative-held seats, Liberal-entry contests

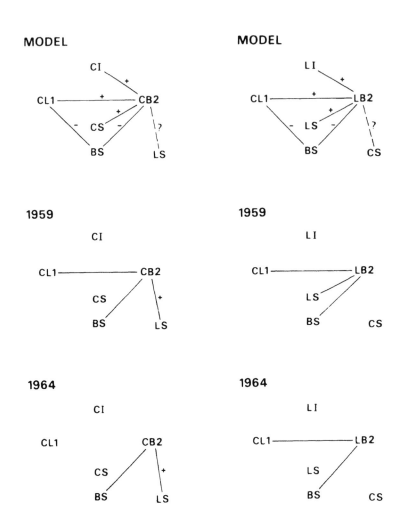

CONSERVATIVE : LIBERAL RATIOS   LABOUR : LIBERAL RATIOS

greater the level of Liberal spending, since the latter party presumably had a good chance of eroding the already-weak Labour base; and (2) the greater the Conservative:Labour ratio at the first election, the smaller the Conservative:Liberal ratio at the second, reflecting the ability of the Liberals to win over more support in the safe than in the marginal Conservative seats. Neither of these hypotheses was substantially validated, however, though the second

Figure 5.13: Significant links in the models for
Labour:Conservative ratios in Labour-held seats, Liberal-entry
contests

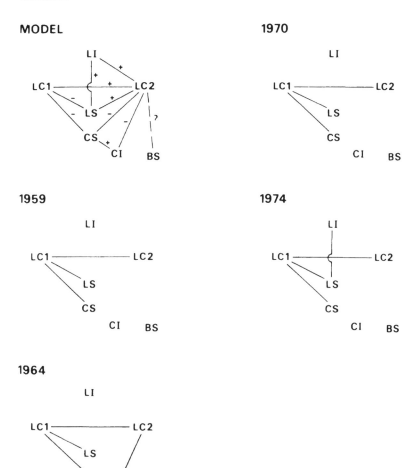

was validated in all four tests in Labour-held seats (Figure
5.12).

With regard to the level of Liberal spending, none of the six
regression coefficients was statistically significant (Figures 5.11–
5.14; Table 5.5); apparently the amount spent by the Liberal party
when fielding a candidate in a constituency not contested at the last
election was in no way related to the electoral situation there, as it

103

Figure 5.14: Significant links in the models for ratios involving
the Liberal party in Labour-held seats, Liberal-entry contests

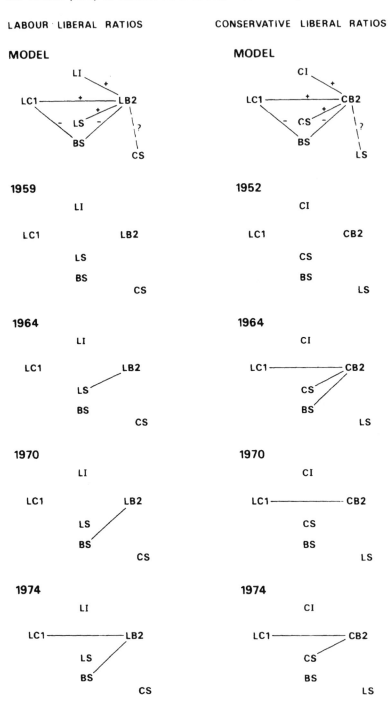

Table 5.5: Summary table of regression results — Liberal-entry contests

	1959	1964	1970	1974
Conservative-held seats				
Regression Coefficients (and R^2) for Spending on Marginality				
CS	x(5)	$-0.075(20)$		
LS	$-0.208(40)$	$-0.275(47)$		
BS	x(1)	x(1)		
Regression Coefficients for CL Ratio on Spending				
CS	x	x		
LS	-0.269	-0.272		
Regression Coefficients for CB Ratio on Spending				
CS	x	x		
BS	-2.813	-1.845		
Labour-held seats				
Regression Coefficients (and R^2) for Spending on Marginality				
LS	$-0.116(21)$	$-0.162(17)$	$-0.043(15)$	$-0.092(21)$
CS	$-0.122(43)$	$-0.145(10)$	$-0.155(62)$	$-0.149(33)$
BS	x(11)	x(1)	x(4)	x(6)
Regression Coefficients for LC Ratio on Spending				
LS	x	x	x	x
CS	x	x	x	x
Regression Coefficients for LB Ratio on Spending				
LS	x	x	x	x
BS	x	x	-8.305	-1.949

x statistically insignificant

is measured in these analyses. With regard to the impact of the previous result, four of the six coefficients were significant, but they were all positive (in Conservative-held seats in 1959 and in Labour-held seats in 1964, 1970 and 1974). The implication is that the greater the strength of the incumbent party over the other 'main party' (Conservative or Labour) at the first election of the pair, the greater its strength relative to the Liberal party at the second; the Liberals could not readily replace a party in a poor second place.

Turning to the impact of the spending, the results with regard to Conservative:Labour contests are quite similar to those for the Two-party type (Table 5.3). In Conservative-held seats, expenditure by the challenger had the expected influence on the results in all four cases (Table 5.5): the more that the challenging party spent, the closer it came to the incumbent party's share of the vote. The incumbent party's spending had no impact on the result, however. In Labour-held seats, the impact was much less (despite the close relationships between marginality and both Conservative and Labour spending). Indeed, neither Labour nor Conservative spending had any impact, and it was only at the last two of the four elections that the amount of Liberal expenditure allowed that party to make significant inroads into the Labour share. With regard to the vote ratios involving the Liberal party, spending by the Liberals significantly influenced its competition with both Conservative and Labour in both years in the Conservative-held seats (Figure 5.12); it had a similar effect in only three of the eight tests in Labour-held seats, however, where overall the models produced very poor fits with no significant links at all in 1959 (Figure 5.14).

THE 1983 ELECTION

The 1983 election differs from the others studied here in three respects. Firstly, all constituencies had Three-party contests. Secondly, the Liberal party contested only approximately half of these; the others were contested by its Alliance partner, the Social Democratic Party, which had only been formed in 1981, which was nationally rather than locally organised (e.g. members' subscriptions were paid directly to national headquarters, which provided much larger sums from central funds for local campaigns), and which had little in the way of local organisation to draw upon in most constituencies, except the Liberal party — which was more cooperative in some places than others — and, in a few, part of the Labour party brought over by 'defectors'. Finally, the election was being held in a set of constituencies different from the previous framework in 1979. Fortunately, as already pointed out (p. 61), the BBC/ITN (1983) calculations of the 1979 results in the 1983 constituencies allowed reasonable estimates of the relevant ratios to be calculated. It was not feasible to decide whether challenger parties had incumbent candidates, however; instead the incumbent variable referred to a candidate who had previously held a seat covering part of the constituency currently being contested.

Previous analyses (Johnston, 1985b, 1986c) have shown substantial differences between the Alliance parties in their spending patterns, so the analyses here relate to four types of contest: each constituency was categorised as Conservative-held or Labour-held and as Liberal- or SDP-contested. The four sets of regression equations are presented in Tables 5.28A–5.31A and are summarised in Figures 5.15–5.17.

Figure 5.15: Significant links in the models for
Conservative:Labour and Labour:Conservative ratios, 1983

CONSERVATIVE - HELD SEATS   LABOUR - HELD SEATS

MODEL   MODEL

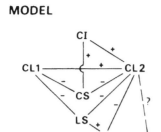   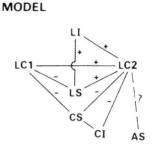

LIBERAL FOR ALLIANCE

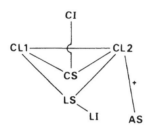   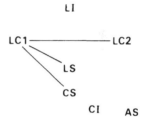

SDP FOR ALLIANCE

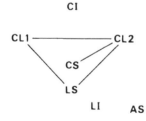   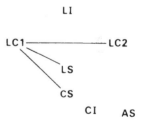

With regard to the pattern of spending, the results for the
Conservative and Labour parties are entirely consistent with those
reported above for all other elections. Both the regression and the
correlation coefficients (relating spending to marginality) were
greater for the challenger than for the incumbent party, though the
former coefficient was insignificant in one case (for Conservative
spending in Conservative-held seats contested by SDP for the
Alliance). Thus, as before, the relationships on the left-hand side of

107

the model (Figure 5.15) were mostly present (though not those for incumbency). For the Alliance parties, the pattern of spending was significantly related to the vote ratio between Liberal and the incumbent party in 1979, though not surprisingly there were closer fits for the regressions involving the Liberal party than for those for the SDP (Figures 5.16 and 5.17). Liberal incumbents raised significantly more than non-incumbents in all four analyses.

Turning to the impact of the spending, there is a marked difference between Conservative and Labour in the effect of their spending on the vote ratios. In the Conservative-held seats, the amount spent by each was significantly related to the electoral outcome; the more that was spent, *ceteris paribus*, the greater the share of the two-party vote. For both parties the regression coefficients were larger in the seats contested by the Liberal party. In addition, the amount spent by the Alliance parties had a significant positive impact on the Conservative:Labour ratio (again, greater in the case of the Liberal-contested constituencies): the more that the Alliance spent, the greater the electoral lead of Conservative over Labour (Alliance candidates came second to Conservative in 70 per cent of the 398 constituencies won by Conservatives, but in only 30 per cent of the 210 won by Labour — Johnston, 1985c. Since they spent more where their ratios to Conservative were smallest, they clearly were able to benefit from this by winning over voters from Labour, leaving the Conservative:Labour ratio higher than it would otherwise have been.)

In the Labour-held seats, by contrast, the amount spent by both Conservative and Labour had no significant impact on the result. Nor did the spending by the Alliance candidates. This suggests that whereas in the Conservative-held seats campaign expenditure was important as an influence in the share of the vote among all three parties, in the Labour-held seats, where in general the Alliance challenge was based on much weaker foundations, expenditure was as irrelevant to the Conservative:Labour share of the vote as at many of the previous elections. Where the Alliance was seeking to push Labour into third place, the amount spent was significant; where it was not, spending was irrelevant to what was closer to a straight Conservative-Labour fight. (All flow-of-the-vote tables show that the Alliance won many more votes from 1979 Labour than Conservative voters — Johnston, 1985a, p. 38 — and estimates of the geography of this flow from Labour to Alliance show that it was much greater in the areas of greatest Conservative strength — Johnston, 1985a, p. 175.)

Figure 5.16: Significant links in the models for incumbent party: Alliance ratios, 1983

CONSERVATIVE - HELD SEATS   LABOUR - HELD SEATS

MODEL

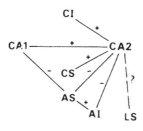
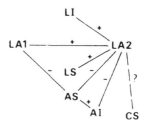

LIBERAL FOR ALLIANCE

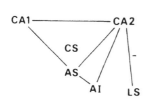
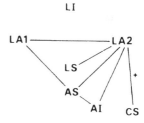

SDP FOR ALLIANCE

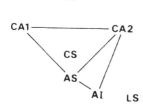
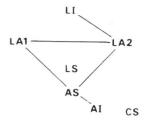

With regard to the vote ratios involving the Alliance parties, Figures 5.16 and 5.17 show that in only one case (the influence of SDP spending on the Labour:Alliance ratio in Conservative-held seats) did the level of Alliance spending fail to have a significant impact on the outcome. In the contests between the Alliance and the party holding a seat, in every case the level of spending by the Alliance was significantly related to the amount spent (as was Alliance incumbency where the Liberal party was involved), and in

Figure 5.17: Significant links in the models for non-incumbent party: Alliance ratios, 1983

CONSERVATIVE-HELD SEATS   LABOUR-HELD SEATS

MODEL

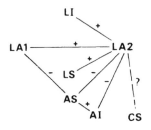
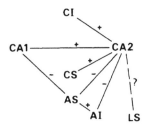

LIBERAL FOR ALLIANCE

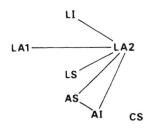
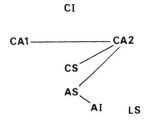

SDP FOR ALLIANCE

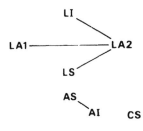
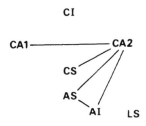

turn the amount spent by the Alliance reduced the Conservative/Labour:Alliance vote ratio. The challenger was able to woo support via campaigning, it seems, but the incumbent party was in general unable to counter this; the exception was Labour spending against a Liberal challenger (Figure 5.16). In addition, in three of the four contests (Figure 5.16), Alliance incumbents had a significant influence on the result, independent of the amount spent; for the first time in these analyses, candidate incumbency comes through as an

110

important influence — for obvious reasons, given the particular context of the 1983 election.

In the contests involving two challenger parties (Figure 5.17), the level of Alliance spending was a significant influence on the outcome in three of the four types; the exception was Conservative-held seats contested by the SDP, where the amount spent by the latter had no significant impact on the Labour:Alliance ratio. (Note that the level of Labour spending did have such an influence, suggesting that Labour campaigning in those seats — in many of which the party came third — was important in establishing a vote-winning credibility.) In all four types, the amount spent by the other challenger was significant, suggesting that overall the contest for second and third place was much affected by the parties' campaign efforts.

Turning finally to the impact of spending by the party not involved in the vote ratio under consideration, Tables 5.28A–5.31A show that Alliance spending significantly benefited the incumbent Conservative party (presumably by winning defectors from Labour) in Conservative-held seats, but a similar impact was not felt in Labour-held seats. The Conservative:Liberal ratio in Conservative-held seats was significantly reduced by Labour spending, whereas the Labour:Liberal ratio in Labour-held seats was significantly increased by Conservative spending.

CHANGES OVER TIME

All of the analyses in this study are cross-sectional, relating to one election only (relative to that preceding it), and no attempt has been made to evaluate trends over more than a pair of adjacent elections. (As noted earlier, this was done in Johnston, 1985d, for the period 1955–70, covering four elections; no longer sequence using the same set of constituencies is possible.) The analyses by Pinto-Duschinsky of the trends in the overall level of spending suggested that its impact should decrease over the period studied here, and three hypotheses were derived in Chapter 3 to represent that deduction. They are evaluated in the present section.

The first of those hypotheses (H_7) suggested that as money became less important to electoral contests, so the relationship between incumbency and spending would decline. Throughout both this chapter and the last, candidate incumbency has been shown as having little impact on either the level of spending or the election result, so the focus here will be on party incumbency only.

The analyses reported in Chapter 4 not only failed to substantiate the hypothesis, but they also indicate a pattern exactly the opposite of that anticipated. The t tests of differences in mean expenditure by Conservative and Labour show that for each, the gap between spending in the constituencies that they held and spending in those in which they were the challenger widened rather than contracted over the period, and was at its greatest in the 1970s. Similarly, the average amounts spent by the Liberal party in Conservative- and Labour-held seats differed much more in the 1970s than in the 1950s. (The Three-party contests provide the clearest statements of this: see p. 63 ff.) Although they were spending less in real terms in 1979 and 1983 than in 1951 and 1959, therefore, the parties varied much more in where they spent it.

The second hypothesis (H_8) suggested, as an extension of the link between incumbency and spending, that over time the link between marginality and spending would decline; as spending became less important, so there would be less effort in raising money in the more marginal constituencies. Once again, the evidence does not substantiate this hypothesis; instead, it suggests that the link has remained consistent throughout the period. This is clear in the summary table (Table 5.2) for Three-party contests, for example.

The third of the hypotheses (H_9) completes the sequence by postulating a decline in the impact of spending on the electoral outcome; less is spent, and so it will have less influence on the electoral fortunes of the parties. Once again, the evidence does not support this interpretation. (As with the evaluation of H_7 and H_8, the Two-party and Liberal-exit contests are of little relevance to this evaluation, since there were none after 1970. The Liberal-entry contests largely refer to the middle years of the study period and so are also of relatively little import. Most of the conclusions must be drawn from the Three-party contests, including 1983.) As Table 5.2 indicates, in the Three-party contests, the probability of a significant regression coefficient between spending and election outcome increased rather than decreased over the period. For the 1951 and 1959 elections, of the sixteen relationships tested, only three were significant: for 1964–70, inclusive, twelve of the 24 regression coefficients were significant; for 1974 and 1979, nine of the sixteen were significant; and in 1983 (Tables 5.28A–5.31A), nine of the sixteen were again significant. (These tabulations exclude the contests involving the two challenger parties — e.g. Conservative

112

and Liberal in Labour-held seats. For them, the significant co-efficients were: 1951–59, 3 of 6; 1964–70, 9 of 12; 1974–79, 7 of 8; and 1983, 7 of 8. Again, there is no evidence of a declining role for spending as an influence on election results.)

None of the three hypotheses has been upheld, therefore. The amount of money spent by the parties in the constituency campaigns has declined in real terms over the period since 1951, but its role in those campaigns — as evaluated by the parties in how much they spend, where it is spent, and as it influenced the results — has not declined: if anything, it has increased.

IN SUMMARY

A full evaluation of the models presented in Chapter 3 is not attempt-ed until Chapter 7, which evaluates the entire set of results. The function of the present summary is to highlight three aspects of the findings reported in the present chapter, and to suggest reasons for them.

The first of these discussions relates to spending and the Liberal party. For much of the period of this study, the Liberals were very much the third party in the English party system, and in the middle of the period reached the nadir of their electoral fortunes during the present century, at which time they could accurately be described as a minor party. In 1950, the Liberal share of the national (i.e. UK) vote was 9.1 per cent (2.6 million votes), and the party contested 475 constituencies of the UK total of 617. One year later, the party's vote had fallen by nearly three-quarters to 730,556 (2.5 per cent of the total), and only 86 constituencies were contested in England — in the few months between the two elections, Liberal candidates 'disappeared' from 323 seats. In terms of percentage of the poll, 1951 was the nadir of the Liberal fortunes, although even less votes were won in 1955. By 1959, the Liberal vote had more than doubled from that figure, and 191 English constituencies had Three-party contests. Five years later, the Liberal vote almost doubled again, with 321 constituencies being contested. The next two elections saw a 25 per cent decline in Liberal support, and in 1970 there were still 288 English constituencies which it failed to contest. The 1974 elec-tions saw a major revival, so that by 1979 virtually all constituencies were being contested — though there was a drop in the Liberal share of the national vote to 13.4 per cent from 18.3 per cent in October 1974. The Alliance, with the advent of the SDP, brought a new surge of support in 1983, to 25.4 per cent of the votes, only 2.2 points less than Labour's share of the poll.

What we see with the Liberal party, therefore, is much less continuity over the period than was the case with either Conservative or Labour. With regard to votes won, the Conservative share varied from 35.8 per cent in October 1974 to 49.4 per cent in 1959; while for Labour, the variation was somewhat greater, from 27.6 per cent in 1983 to 48.8 per cent in 1951. Most importantly for the present study, however, both Conservative and Labour contested virtually every seat in England over the period, whereas the Liberals contested a minority at certain elections. (There were 139 constituencies that provided Three-party contests at all of the 1959, 1964, 1966 and 1970 elections, thereby having some continuity of Liberal candidacy, out of 513.)

Given this absence of continuity in Liberal candidacy, together with its relatively poor performance over much of the period, one would anticipate that expenditure on the constituency campaigns should have a substantial effect; a poorly-placed party, with a weak organisational base in many areas, little publicity in the national media for its policies and major figures and, very largely, constituency candidates with little or no political experience (including local government experience in many cases) should benefit from spending on publicising its candidates to an electorate which knew little of the Liberals but was prepared, it seems, to use their party as a focus for protest votes (see Himmelweit *et al.*, 1985) if informed about its candidates. In the Three-party contests, Liberals clearly raised more in the more marginal (for them) constituencies and benefited from their spending in its contests with both Conservative and, from 1966 onwards, Labour incumbents. In the Liberal-entry contests, their level of spending was not related to the potential political benefits (the relative shares of the vote at the previous election won by the Conservatives and Labour), but what was spent had some influence in winning over voters from the other two parties. In 1983, Liberal spending was very influential on the result.

Throughout the analyses the Liberal party (and the Alliance in 1983) has been presented as a challenger; the number of seats that it was defending was never sufficient to justify separate analyses. The amount of money that it was able to raise invariably meant that it was the poorest of the three parties, as Table 4.2 shows. However, the impact of its spending was much greater than that of the Conservative or Labour parties: in the Three-party contests, for example, the regression coefficients for BS are much larger than those for CS and LS (Table 5.2), as they are also in the Liberal-entry contests (Table 5.5). In relative terms, therefore, Liberal spending on publicity and information/advertising brought much greater benefits in terms of share of the vote than did Conservative or Labour spending: this was also generally the case in 1983. Thus, the greater the relative voter ignorance, the more it can be countered by campaign spending intended to inform.

The second discussion point to be raised here relates to the differences in results between the Three-party and the Two-party contests with regard to the impact of Conservative and Labour spending. In general, as Tables 5.2 and 5.3 indicate, spending by these two parties had a greater impact on their relative shares of the vote in the former than in the latter. One reason suggested for this was that the absence of the Liberal party meant that there was more likely to be a stable set of preferences in the Two-party contests that were not as amenable in terms of advertising and information. This argument is difficult to sustain, however, because not all of the Two-party contests were in the same category for a sequence of elections, and the presence of the Liberal candidates does not readily account for the impact of Labour-challenger spending on Conservative incumbents and vice versa.

One possible reason not yet explored relates to the nature of the type of seat involved, not just with reference to the electoral contest but also to its socio-economic characteristics. This possibility was explored in a number of analyses in which constituencies were categorised according to a range of characteristics for which data were available from the 1981 Census. These produced very few significant differences, and the only important ones are summarised in Table 5.6. In the Conservative-held seats, Labour spent more in the non-agricultural than in the agricultural constituencies and more in the manufacturing than the non-manufacturing — not surprisingly, since the manufacturing/non-agricultural populations have traditionally provided the bulk of its support; and also not surprisingly, Alliance spending showed the inverse of that pattern. In the Labour-held seats, both Conservative and the Alliance parties spent more, unsurprisingly, in the non-mining seats, and the Alliance in the non-manufacturing seats also. Labour also spent less in the mining than in the non-mining seats, and yet the coalfield constituencies have traditionally provided Labour's safest seats.

The last finding in the preceding paragraph is important; presumably Labour spent less in the mining seats because they were so safe. Moreover, few mining seats had Three-party contests in the

Table 5.6: Average spending by type of constituency in 1983[a]

	Average spending by		
	Conservative	Labour	Alliance
Conservative-won seats in 1979			
Agricultural	86.6	46.4	69.4
Non-agricultural	84.7	61.6	61.0
Manufacturing	84.4	71.8	52.0
Non-manufacturing	85.2	56.9	64.1
Labour-won seats in 1979			
Mining	50.3	60.0	31.7
Non-mining	62.9	80.0	51.6
Manufacturing	67.1	77.5	41.7
Non-manufacturing	57.8	76.0	50.9

Note a: Agricultural constituencies are those with 5 per cent or more of the workforce employed in agriculture; manufacturing constituencies are those with 36 per cent or more so employed; mining constituencies are those with at least 7 per cent employed in energy industries.

early elections of the sequence studied here. In general, therefore, one might suggest that if the pattern for 1983 were typical of the whole period, then spending had less impact in the Two-party contests because the seats were more safe; party lines were clearly drawn and very few people were likely to be influenced by spending campaigns. This is explored further by looking at the Conservative:Labour and Labour:Conservative ratios for the various types of contest at the different elections up to 1974 (Table 5.7).

With regard to the Conservative-held seats, there is a clear and significant difference between the Three-party and the Two-party contests. At each of the elections (1951–70), the Conservative: Labour ratios were significantly higher in the Three-party than in the Two-party contests, indicating that the former were on average much safer seats. (Interestingly, the Liberal-exit ratios were closer to the latter, suggesting that Liberal withdrawals were from the more marginal seats where less 'protest vote' transfers could be expected: Johnston and Hay, 1983.) In the Labour-held seats, on the other hand, this distinction is largely absent, and certainly so after 1959 when the Two-party contests are the safer seats. There is, then, no consistent interpretation for these findings provided by analyses of the nature of the constituencies and the contests in them.

The most viable interpretation is linked to the third point to be highlighted here, which concerns the failure of the hypotheses relating to trends over time. The influence of spending on election results has increased, not decreased, and it was clearly an important

Table 5.7: The average Conservative:Labour and Labour:Conservative ratios, by type of contest

	Conservative-held CL				Labour-held LC			
	First election		Second election		First election		Second election	
	Mean	SD	Mean	SD	Mean	SD	Mean	SD
1951								
Three-party	1.719	0.53	1.779	0.55	2.959	0.33	2.419	0.20
Two-party	1.561	0.42	1.507	0.35	1.939	0.88	1.967	0.89
Liberal-exit	1.562	0.45	1.539	0.48	1.491	0.64	1.670	0.76
1959								
Three-party	1.856	0.71	1.998	0.74	1.765	0.90	1.701	0.86
Two-party	1.631	0.47	1.539	0.46	1.516	0.64	1.625	0.68
Liberal-entry	2.023	0 86	1.724	0.64	1.444	0.74	1.443	0.84
1964								
Three-party	2.009	0.81	1.710	0.73	1.639	0.93	1.847	1.09
Two-party	1.397	0.43	1.174	0.33	1.673	0.72	1.885	0.75
Liberal-entry	1.622	0.47	1.472	0.49	1.379	0.42	1.774	0.66
1966								
Three-party	1.761	0.64	1.513	0.62	1.615	0.82	1.855	1.03
Two-party	1.249	0.28	1.126	0.23	1.758	0.75	2.121	0.96
Liberal-exit	1.368	0.36	1.121	0.22	1.535	0.30	1.601	0.60
Liberal-entry	1.407	0.37	1.235	0.37	1.409	0.30	1.949	0.74
1970								
Three-party	1.550	0.47	1.935	0.57	1.601	0.71	1.234	0.45
Two-party	1.233	0.18	1.486	0.24	2.002	0.92	1.618	0.70
Liberal-exit	1.554	0.47	1.341	0.54	1.782	1.04	1.321	0.56
Liberal-entry	1.255	0.21	1.688	0.40	1.682	0.87	1.426	0.77
1974								
Three-party	1.967	0.91	1.726	0.69	1.747	0.79	2.030	1.02
Liberal-entry	-		-		2.013	0.84	2.661	1.35

factor in 1983. (See Johnston, 1986e, on how important, and p. 192).
Why should this be so, when in real terms less has been spent? The
answer appears to lie in Table 4.1, which relates the amount spent
to the amount allowed, and in Tables 4.2–4.6, which summarise the
volume of spending relative to the maxima. For 1951 to 1966 the
limits on spending were unaltered. This was a period of declining
value of money, and it was easier for the parties to raise the
maximum or close to it. (The amount spent in 1966 was only 19 per
cent up on the 1951 total. In real terms, however, this was a substantial decline; Pinto-Duschinsky (1981a) uses the cost-of-living index
to suggest a 56 per cent decline in the value of the £ between the two
dates.) In particular, until 1966 the Conservative party was able to
achieve an average expenditure of over 90 per cent in the seats which
it held, with small standard deviations, and spending of over 75 per
cent (in many cases substantially so) in Labour-held seats. Since
1966, Conservative spending has both declined on average and
become spatially more variable. The same is true of Labour spending, from slightly lower peaks in the 1950s; and Liberal spending
declined very substantially, especially in the Labour-held seats.

When spending is high and consistent across the constituencies,
it is less likely to be related to marginality or to have an impact on
the election outcome. When it is variable from constituency to
constituency, then it is more likely to be linked to marginality and
to influence the result. The degree of variability increased from 1970
onwards as a consequence of the relaxation of the limits, with consequences as just indicated.

Thus, the increase in the limits from 1970 onwards created the
situation in which, because of the inability (or unwillingness,
because they perceived it as unnecessary) of the parties to raise all
that was allowed in every constituency, spending was more likely to
have an impact. Spending has been more important as an influence
in recent years because of both the differential ability of the parties,
and of the constituency organisations within each party, to raise the
allowed sums.

This interpretation not only accounts for the increasing importance of the spending activities of constituency parties over time, but
also provides an explanation for the difference between the Three-
party and the Two-party contests. Most of the latter occurred in the
period up to 1970 (as also did most of the Liberal-exit contests),
when spending was relatively invariant. Since 1970, Three-party
contests have been the norm, spending has been more variable, and
more votes have gone to the constituency parties which spent more.

118

Table 5.1A: Regression results, 1951 — Three-party contests, Conservative-held seats

Dependent	Independent	a	b	t	beta	R^2
CS		0.964				
	CL50		−0.014	0.62	−0.11	
	CI		0.043	1.30	0.24	0.14
LS		1.152				
	CL50		−0.241	3.34	−0.66	
	LI		0.115	2.07	0.30	0.45
CL51		0.369				
	CL50		0.963	16.13	0.92	
	CS		−0.058	0.16	−0.01	
	LS		−0.252	1.48	−0.09	
	BS		−0.042	0.37	−0.02	
	CI		0.042	0.67	0.28	
	LI		−0.025	0.48	−0.02	0.95
BS		0.925				
	CB50		−0.114	2.61	−0.60	
	LB50		−0.030	0.60	−0.05	
	BI		0.044	0.63	0.09	0.45
CB5		1.700				
	CB50		1.051	5.39	0.63	
	CS		0.565	0.21	0.02	
	BS		−3.367	3.49	−0.38	
	LS		1.737	1.83	0.17	
	CI		0.272	0.57	0.05	
	BI		−0.360	0.92	−0.09	0.82
LB51		1.179				
	LB50		1.422	6.60	0.78	
	LS		−0.298	0.31	−0.03	
	BS		−0.641	0.87	−0.08	
	CS		0.743	0.36	0.02	
	LI		0.175	0.61	0.05	
	BI		−0.356	1.19	−0.10	0.89

119

Table 5.2A: Regression results, 1951 — Three-party contests, Labour-held seats

Dependent	Independent	a	b	t	beta	R^2
LS		0.583				
	LC50		0.001	0.58	0.23	
	LI		0.062	0.14	0.15	0.29
CS		0.953				
	LC50		− 0.001	0.48	− 0.18	
	CI		0.001	0.07	0.02	0.44
LC51		2.235				
	LC50		0.653	6.14	1.07	
	LS		0.119	0.05	0.01	
	CS		0.529	0.30	0.04	
	BS		− 3.143	2.45	− 0.31	
	LI		− 0.611	0.79	− 0.08	
	CI		− 0.445	0.93	− 0.11	0.90
BS		0.615				
	LB50		0.001	0.01	0.15	
	CB50		− 0.059	1.19	− 0.50	
	BI		0.086	0.56	0.23	0.21
LB51		0.957				
	LB50		0.879	5.50	0.91	
	LS		1.767	0.41	0.06	
	BS		− 1.563	0.57	− 0.09	
	CS		0.593	0.21	0.03	
	LI		0.115	0.08	0.01	
	BI		0.053	0.06	0.01	0.86
CB51		3.400				
	CB50		1.764	4.63	1.11	
	CS		5.546	1.59	0.34	
	BS		− 4.815	1.96	− 0.36	
	LS		1.548	0.40	0.07	
	CI		0.403	0.50	0.08	
	BI		− 1.426	1.38	− 0.28	0.83

Table 5.3A: Regression results, 1959 — Three-party contests, Conservative-held seats

Dependent	Independent	a	b	t	beta	R^2
CS		1.104				
	CL55		−0.033	1.72	−0.24	
	CI		−0.026	0.89	−0.14	0.09
LS		1.022				
	CL55		−0.158	3.45	−0.51	
	LI		−0.035	0.65	−0.08	0.44
CL59		0.265				
	CL55		0.967	19.08	0.92	
	CS		−0.045	0.15	−0.00	
	LS		−0.165	1.04	−0.05	
	BS		0.146	1.12	0.04	
	CI		0.021	0.34	0.01	
	LI		0.001	0.02	0.00	0.93
BS		1.066				
	CB55		−0.116	4.53	−0.78	
	LB55		0.011	0.44	0.07	
	BI		0.001	0.07	0.00	0.49
CB59		1.787				
	CB55		0.436	8.07	0.65	
	CS		−0.148	0.23	−0.01	
	BS		−1.259	3.52	−0.28	
	LS		0.508	1.89	0.11	
	CI		0.040	0.30	0.02	
	BI		−0.184	0.86	−0.05	0.82
LB59		0.699				
	LB55		0.445	8.17	0.67	
	LS		0.610	2.25	0.15	
	BS		−0.698	2.71	−0.16	
	CS		0.191	0.39	0.02	
	LI		−0.077	0.74	−0.04	
	BI		−0.202	1.23	−0.06	0.88

Table 5.4A: Regression results, 1964 — Three-party contests, Conservative-held seats

Dependent	Independent	a	b	t	beta	R^2
CS		1.030				
	CL59		−0.000	0.65	−0.05	
	CI		−0.033	1.65	−0.14	0.03
LS		1.120				
	CL59		−0.134	6.47	−0.55	
	LI		−0.001	0.19	−0.01	0.43
CL64		−0.195				
	CL59		0.770	21.11	0.85	
	CS		0.367	1.59	0.05	
	LS		−0.387	2.56	−0.11	
	BS		0.359	3.17	0.10	
	CI		0.037	0.68	0.02	
	LI		0.092	1.15	0.03	0.87
BS		1.049				
	CB59		−0.059	2.97	−0.34	
	LB59		−0.020	0.98	−0.11	
	BI		−0.028	0.83	−0.06	0.15
CB64		1.538				
	CB59		0.412	11.21	0.58	
	CS		−0.086	0.22	−0.01	
	BS		−1.515	7.34	−0.37	
	LS		0.903	4.68	0.22	
	CI		0.054	0.59	0.03	
	BI		0.042	0.50	0.02	0.69
LB64		1.254				
	LB59		0.504	8.84	0.54	
	LS		1.381	4.79	0.27	
	BS		−1.530	6.66	−0.29	
	CS		−0.318	0.73	−0.03	
	LI		−0.026	0.18	−0.01	
	BI		0.043	0.47	0.02	0.76

122

Table 5.5A: Regression results, 1964 — Three-party contests, Labour-held seats

Dependent	Independent	a	b	t	beta	R^2
LS		1.151				
	LC59		−0.088	1.86	−0.43	
	LI		−0.087	0.76	−0.15	0.14
CS		0.991				
	LC59		−0.136	2.85	−0.55	
	CI		−0.197	2.29	−0.37	0.36
LC64		0.609				
	LC59		1.055	29.84	0.91	
	LS		−0.121	0.65	−0.02	
	CS		−0.448	2.66	−0.09	
	BS		−0.163	1.43	−0.04	
	LI		0.096	1.21	0.03	
	CI		−0.032	0.48	−0.01	0.99
BS		1.122				
	LB59		−0.165	4.17	−0.71	
	CB59		0.031	0.65	0.11	
	BI		−0.043	0.37	−0.06	0.37
LB64		0.322				
	LB59		0.723	4.92	0.81	
	LS		0.356	0.39	0.07	
	BS		−0.429	0.61	−0.11	
	CS		0.012	0.02	0.01	
	LI		0.327	0.85	0.10	
	BI		−0.062	0.18	−0.02	0.67
CB64		−0.343				
	CB59		0.711	4.16	0.75	
	CS		0.442	0.60	0.11	
	BS		−0.044	0.07	−0.01	
	LS		0.316	0.36	0.07	
	CI		0.205	0.76	0.10	
	BI		−0.116	0.37	−0.05	0.64

123

Table 5.6A: Regression results, 1966 — Three-party contests, Conservative-held seats

Dependent	Independent	a	b	t	beta	R^2
CS		0.932				
	CL64		−0.016	1.85	−0.14	
	CI		−0.001	0.39	0.03	0.03
LS		0.802				
	CL64		−0.110	4.39	−0.37	
	LI		0.001	0.26	0.01	0.39
CL66		−0.616				
	CL64		0.908	28.84	0.94	
	CS		0.385	1.67	0.04	
	LS		0.082	0.80	0.02	
	BS		0.212	2.72	0.07	
	CI		0.001	0.12	0.00	
	LI		0.020	0.36	0.00	0.88
BS		0.790				
	CB64		−0.077	2.44	−0.27	
	LB64		−0.041	1.41	−0.16	
	BI		0.073	1.95	0.13	0.20
CB66		−0.331				
	CB64		0.935	16.71	0.70	
	CS		0.645	1.16	0.05	
	BS		−0.843	4.29	−0.17	
	BS		1.235	5.98	0.23	
	CI		−0.042	0.46	−0.02	
	BI		0.023	0.24	0.00	0.76
LB66		−0.105				
	LB64		1.167	15.34	0.74	
	LS		0.730	2.74	0.11	
	BS		−0.601	2.89	−0.10	
	LS		0.651	1.12	0.04	
	LI		0.000	0.00	0.00	
	BI		0.066	0.64	0.02	0.84

Table 5.7A: Regression results, 1966 — Three-party contests, Labour-held seats

Dependent	Independent	a	b	t	beta	R^2
LS		1.029				
	LC64		−0.078	3.18	−0.39	
	LI		−0.091	2.00	−0.27	0.33
CS		1.080				
	LC64		−0.131	5.48	−0.60	
	CI		−0.054	1.33	−0.15	0.49
LC66		0.146				
	LC64		1.166	14.96	0.93	
	LS		0.341	0.92	0.05	
	CS		−0.573	1.50	−0.10	
	BS		0.187	0.85	0.04	
	LI		−0.128	1.04	−0.06	
	CI		−0.041	0.38	−0.02	0.88
BS		0.539				
	LB64		−0.001	0.29	−0.08	
	CB64		0.030	0.79	0.24	
	BI		0.073	0.80	0.10	0.13
LB66		4.633				
	LB64		0.455	2.60	0.32	
	LS		0.940	0.35	0.05	
	BS		−2.590	1.58	−0.23	
	CS		−2.374	0.98	−0.15	
	LI		0.627	0.69	0.11	
	BI		0.189	0.17	0.02	0.26
CB66		2.316				
	CB64		0.198	2.10	0.24	
	CS		0.438	0.36	0.05	
	BS		−2.036	2.91	−0.32	
	LS		1.224	1.06	0.14	
	CI		0.947	2.67	0.32	
	BI		0.083	0.18	0.02	0.52

Table 5.8A: Regression results, 1970 — Three-party contests, Conservative-held seats

Dependent	Independent	a	b	t	beta	R^2
CS		0.902				
	CL66		−0.034	1.75	−0.15	
	CI		−0.001	0.12	−0.00	0.03
LS		0.639				
	CL66		−0.117	3.52	−0.32	
	LI		0.056	1.45	0.09	0.39
CL70		0.125				
	CL66		1.076	28.65	0.89	
	CS		0.325	2.22	0.06	
	LS		−0.341	3.30	−0.10	
	BS		0.149	1.77	0.05	
	CI		0.001	0.23	0.01	
	LI		−0.019	0.37	−0.01	0.90
BS		0.705				
	CB66		−0.109	3.71	−0.55	
	LB66		0.019	0.74	1.11	
	BI		0.072	2.10	0.15	0.25
CB70		2.398				
	CB66		0.757	8.52	0.55	
	CS		−0.592	0.82	−0.05	
	BS		−1.920	4.44	−0.28	
	LS		1.146	2.69	0.15	
	CI		0.067	0.43	0.02	
	BI		0.214	1.20	0.06	0.59
LB70		1.702				
	LB66		0.634	9.51	0.60	
	LS		1.169	3.68	0.19	
	BS		−1.157	4.30	−0.19	
	CS		−0.597	1.40	−0.06	
	LI		−0.088	0.60	−0.02	
	BI		0.123	1.11	0.04	0.78

Table 5.9A: Regression results, 1970 — Three-party contests, Labour-held seats

Dependent	Independent	a	b	t	beta	R^2
LS		0.989				
	LC66		-0.103	3.13	-0.45	
	LI		-0.082	1.49	-0.25	0.36
CS		1.098				
	LC66		-0.186	6.00	-0.75	
	CI		0.057	1.26	0.16	0.55
LC70		0.669				
	LC66		0.534	12.30	0.84	
	LS		0.048	0.35	0.02	
	CS		-0.443	2.83	-0.17	
	BS		-0.037	0.42	-0.02	
	LI		0.047	0.91	0.05	
	CI		0.047	1.04	0.05	0.92
BS		0.608				
	LB66		0.001	0.43	0.09	
	CB66		-0.096	2.74	-0.65	
	BI		0.046	0.58	0.07	0.26
LB70		3.219				
	LB66		2.423	2.00	0.21	
	BS		-4.131	5.15	-0.47	
	CS		-0.179	0.16	-0.02	
	LI		0.050	0.11	0.01	
	BI		0.040	0.09	0.01	0.68
CB70		1.312				
	CB66		0.272	2.36	0.23	
	CS		2.593	2.45	0.27	
	BS		-4.221	6.19	-0.54	
	LS		2.424	2.62	0.23	
	CI		0.181	0.56	0.05	
	BI		-0.146	0.42	-0.03	0.79

Table 5.10A: Regression results, 1974 — Three-party contests, Conservative-held seats

Dependent	Independent	a	b	t	beta	R^2
CS		1.006				
	CL4F		−0.017	2.34	−0.15	
	CI		0.015	0.93	0.06	0.05
LS		0.746				
	CL4F		−0.093	5.88	−0.41	
	LI		0.077	2.63	0.14	0.44
CL40		0.477				
	CL4F		0.675	41.25	0.89	
	CS		0.132	1.13	0.02	
	LS		−0.289	3.91	−0.09	
	BS		0.072	1.40	0.02	
	CI		−0.042	1.51	−0.02	
	LI		−0.106	3.23	−0.06	0.94
BS		1.010				
	CB4F		−0.166	4.59	−0.34	
	LB4F		−0.105	2.99	−0.24	
	BI		0.026	0.85	0.04	0.30
CB40		0.880				
	CB4F		0.571	10.72	0.47	
	CS		0.204	0.86	0.04	
	BS		−0.813	7.23	−0.32	
	LS		0.670	5.53	0.23	
	CI		0.070	1.22	0.05	
	BI		−0.135	2.45	−0.09	0.64
LB40		0.242				
	LB4F		0.806	12.78	0.57	
	LS		0.728	5.31	0.20	
	BS		−0.535	5.28	−0.16	
	CS		0.283	1.39	0.04	
	LI		0.283	4.87	1.34	
	BI		−0.111	2.34	−0.06	0.86

128

Table 5.11A: Regression results, 1974 — Three-party contests, Labour-held seats

Dependent	Independent	a	b	t	beta	R^2
LS		0.897				
	LC4F		−0.100	6.61	−0.46	
	LI		0.063	2.16	0.14	0.20
CS		1.033				
	LC4F		−0.191	12.35	−0.68	
	CI		0.016	0.47	0.02	0.48
LC40		0.081				
	LC4F		1.207	41.83	0.94	
	LC		−0.022	0.20	−0.01	
	CS		−0.300	2.75	−0.06	
	BS		0.207	2.23	0.04	
	LI		0.015	0.35	0.00	
	CI		−0.044	0.95	−0.02	0.96
BS		0.363				
	LB4F		−0.057	4.30	−0.82	
	CB4F		0.070	4.09	0.78	
	BI		0.076	2.15	0.15	0.12
LB40		6.017				
	LB4F		0.154	4.72	0.28	
	LS		−0.651	1.08	−0.08	
	BS		−1.028	−3.83	−0.25	
	CS		−2.170	4.46	−0.33	
	LI		0.095	0.42	0.03	
	BI		−0.167	0.71	−0.04	0.36
CB40		1.972				
	CB4F		0.051	2.69	0.15	
	CS		0.837	3.25	0.26	
	BS		−1.103	5.04	−0.31	
	LS		0.265	1.00	0.07	
	CI		0.174	1.57	0.92	
	BI		−0.170	1.64	−0.09	0.43

129

Table 5.12A: Regression results, 1979 — Three-party contests, Conservative-held seats

Dependent	Independent	a	b	t	beta	R^2
CS		0.909				
	CL40		−0.021	1.76	−0.13	
	CI		0.017	0.79	0.05	0.02
LS		0.420				
	CL40		−0.067	2.78	−0.21	
	LI		0.027	0.75	0.04	0.41
CL79		−0.645				
	CL40		1.369	25.14	0.81	
	CS		0.519	2.03	0.05	
	LS		−0.280	1.70	−0.05	
	BS		0.855	6.04	0.17	
	CI		−0.011	0.13	−0.00	
	LI		−0.061	0.65	−0.02	0.85
BS		0.890				
	CB40		−0.214	5.35	−0.53	
	LB40		−0.001	0.07	−0.00	
	BI		0.057	1.88	0.10	0.29
CB79		1.140				
	CB40		1.093	12.93	0.55	
	CS		0.409	1.26	0.04	
	BS		−1.582	7.94	−0.32	
	LS		0.487	2.60	0.10	
	CI		0.040	0.37	0.01	
	BI		−0.259	2.80	−0.09	0.73
LB79		0.088				
	LB40		1.103	17.25	0.75	
	LS		0.459	3.12	0.10	
	BS		−0.745	5.86	−0.16	
	CS		0.184	0.87	0.02	
	LI		0.306	3.95	0.09	
	BI		−0.133	2.27	−0.05	0.88

Table 5.13A: Regression results, 1979 — Three-party contests, Labour-held seats

Dependent	Independent	a	b	t	beta	R^2
LS		0.875				
	LC40		−0.084	6.69	−0.51	
	LI		0.024	0.86	0.05	0.23
CS		0.721				
	LC40		−0.096	7.93	−0.48	
	CI		−0.020	0.63	−0.03	0.47
LC79		0.902				
	LC40		0.452	18.65	0.78	
	LS		−0.119	0.91	−0.03	
	CS		−0.306	2.48	−0.11	
	BS		−0.261	1.83	−0.06	
	LI		0.001	0.02	0.00	
	CI		0.001	0.07	0.00	0.78
BS		0.280				
	LB40		−0.031	4.58	−0.28	
	CB40		0.020	1.37	0.09	
	BI		0.051	2.00	0.13	0.11
LB79		4.039				
	LB40		0.560	8.25	0.45	
	LS		0.391	0.74	0.04	
	BS		−4.336	7.26	−0.38	
	CS		−0.224	0.50	−0.03	
	LI		0.056	0.25	0.01	
	BI		−0.384	1.67	−0.08	0.44
CB79		2.400				
	CB40		0.802	6.06	0.40	
	CS		0.799	2.00	0.14	
	BS		−2.074	4.60	−0.23	
	LS		0.367	0.89	0.05	
	CI		0.352	1.86	0.09	
	BI		−0.214	1.20	−0.06	0.46

Table 5.14A: Regression results, 1951 — Two-party contests

Dependent	Independent	a	b	t	beta	R^2
Conservative-held seats						
CS		1.108				
	CL50		−0.028	0.51	−0.11	
	CI		−0.146	1.55	−0.32	0.14
LS		1.361				
	CL50		−0.447	5.42	−0.77	
	LI		−0.001	0.03	−0.01	0.59
CL51		−0.058				
	CL50		1.137	19.74	0.95	
	CS		−0.086	0.57	−0.02	
	LS		−0.120	1.22	−0.06	
	CI		0.085	1.28	0.04	
	LI		−0.029	0.97	−0.03	0.98
Labour-held seats						
LS		0.857				
	LC50		−0.065	3.20	−0.38	
	LI		0.088	1.28	0.15	0.18
CS		1.104				
	LC50		−0.145	9.77	0.78	
	CI		0.001	0.23	0.02	0.62
LC51		0.110				
	LC50		0.966	49.43	0.98	
	LS		−0.001	0.08	−0.01	
	CS		−0.074	0.72	−0.01	
	LI		0.005	0.13	0.01	
	CI		−0.028	1.19	−0.01	0.99

Table 5.15A: Regression results, 1959 — Two-party contests

Dependent	Independent	a	b	t	beta	R^2
Conservative-held seats						
CS		1.145				
	CL55		−0.104	6.32	−0.48	
	CI		−0.038	2.07	−0.19	0.23
LS		1.358				
	CL55		−0.362	13.26	−0.75	
	LI		0.027	0.85	0.06	0.58
CL59		0.530				
	CL55		0.926	26.17	0.92	
	CS		−0.023	0.33	−0.01	
	LS		−0.328	2.73	−0.07	
	CI		−0.001	0.39	−0.01	
	LI		−0.012	0.44	−0.01	0.94
Labour-held seats						
LS		1.028				
	LC55		−0.095	4.67	−0.35	
	LI		−0.017	0.55	−0.05	0.11
CS		1.089				
	LC55		−0.169	10.07	−0.62	
	CI		0.021	0.79	0.06	0.38
LC59		0.066				
	LC55		0.896	41.56	0.95	
	LS		0.272	4.10	0.08	
	CS		−0.281	3.49	−0.08	
	LI		−0.024	0.93	−0.02	
	CI		0.003	0.01	0.00	0.95

133

Table 5.16A: Regression results, 1964 — Two-party contests

Dependent	Independent	a	b	t	beta	R^2
Conservative-held seats						
CS		1.069				
	CL59		−0.052	3.10	−0.41	
	CI		0.001	0.22	0.03	0.11
LS		1.359				
	CL59		−0.338	8.09	−0.75	
	LI		−0.055	0.88	−0.08	0.57
CL64		0.001				
	CL59		0.677	12.55	0.89	
	CS		0.442	1.55	0.07	
	LS		−0.231	2.01	−0.14	
	CI		0.001	0.04	0.01	
	LI		−0.133	2.57	−0.12	0.90
Labour-held seats						
LS		1.069				
	LC59		−0.052	2.91	−0.26	
	LI		−0.066	1.85	−0.17	0.07
CS		1.072				
	LC59		−0.143	7.84	−0.60	
	CI		0.001	0.23	0.02	0.34
LC64		0.170				
	LC59		0.999	31.89	0.95	
	LS		0.181	1.56	0.04	
	CS		−0.181	1.32	−0.04	
	LI		0.038	0.89	0.02	
	CI		−0.023	0.51	−0.01	0.93

134

Table 5.17A: Regression results, 1966 — Two-party contests

Dependent	Independent	a	b	t	beta	R^2
Conservative-held seats						
CS		1.086				
	CL64		−0.110	2.18	−0.46	
	CI		−0.019	0.50	−0.13	0.15
LS		1.676				
	CL64		−0.604	6.64	−0.88	
	LI		−0.154	2.07	−0.34	0.67
CL66		0.026				
	CL64		0.842	8.22	0.98	
	CS		0.068	0.25	0.02	
	LS		0.025	0.17	0.02	
	CI		−0.082	1.87	−0.15	
	LI		−0.081	1.56	−0.14	0.92
Labour-held seats						
LS		0.941				
	LC64		−0.034	2.09	−0.17	
	LI		−0.083	3.29	−0.27	0.12
CS		0.993				
	LC64		−0.140	7.97	−0.55	
	CI		0.075	2.61	0.18	0.42
LC66		−0.105				
	LC64		1.264	36.37	0.98	
	LS		0.001	0.05	0.01	
	CS		0.074	0.51	0.01	
	LI		−0.065	1.43	−0.03	
	CI		−0.053	1.10	−0.03	0.94

135

Table 5.18A: Regression results, 1970 — Two-party contests

Dependent	Independent	a	b	t	beta	R^2
Conservative-held seats						
CS		1.242				
	CL66		−0.219	2.12	−0.42	
	CI		−0.098	1.85	−0.40	0.23
LS		1.454				
	CL66		−0.578	3.51	−0.55	
	LI		0.140	0.89	0.16	0.51
CL70		1.118				
	CL66		0.949	4.22	0.72	
	CS		0.046	0.18	0.04	
	LS		−0.843	2.10	−0.33	
	CI		−0.098	0.99	−0.16	
	LI		−0.001	0.05	−0.00	0.67
Labour-held seats						
LS		0.932				
	LC66		−0.073	4.58	−0.36	
	LI		−0.083	2.89	−0.22	0.22
CS		1.023				
	LC66		−0.163	12.23	−0.70	
	CI		0.039	1.47	0.08	0.59
LC70		0.342				
	LC66		0.722	33.84	0.95	
	LS		−0.061	0.76	−0.02	
	CS		−0.132	1.38	−0.04	
	LI		−0.070	2.50	−0.05	
	CI		0.020	0.69	0.01	0.95

Table 5.19A: Regression results, 1951 — Liberal-exit contests

Dependent	Independent	a	b	t	beta	R^2
Conservative-held seats						
CS		1.029				
	CL50		-0.056	4.38	-0.32	
	CI		0.001	0.30	0.02	0.10
LS		1.026				
	CL50		-0.213	8.22	-0.54	
	LI		-0.016	0.57	-0.04	0.29
CL51		0.085				
	CL50		0.935	70.34	0.99	
	CS		0.113	1.69	0.02	
	LS		-0.013	0.38	0.01	
	CI		-0.059	2.51	-0.03	
	LI		-0.016	1.39	-0.03	0.98
Labour-held seats						
LS		1.105				
	LC50		-0.111	7.32	-0.52	
	LI		-0.084	1.63	-0.11	0.27
CS		1.037				
	LC50		-0.067	6.38	-0.46	
	CI		-0.011	0.61	-0.04	0.21
LC51		0.192				
	LC50		0.823	53.57	0.97	
	LS		-0.087	1.30	-0.02	
	CS		-0.018	0.19	-0.01	
	LI		0.012	0.28	0.01	
	CI		0.014	0.67	0.01	0.97

Table 5.20A: Regression results, 1966 — Liberal-exit contests

Dependent	Independent	a	b	t	beta	R^2
Conservative-held seats						
CS		0.996				
	CL64		−0.072	2.69	−0.41	
	CI		0.041	1.66	0.33	0.20
LS		1.228				
	C64		−0.325	4.67	−0.64	
	LI		0.000	0.15	0.03	0.37
CL66		0.222				
	CL64		0.603	13.78	0.95	
	CS		0.166	0.73	0.04	
	LS		−0.062	0.70	−0.04	
	CI		−0.028	0.86	−0.06	
	LI		−0.041	1.28	−0.09	0.91
Labour-held seats						
LS		1.068				
	LC64		−0.133	4.58	−0.63	
	LI		0.001	0.02	0.00	0.35
CS		1.123				
	LC64		−0.196	7.00	−0.78	
	CI		−0.001	0.12	−0.01	0.57
LC66		1.019				
	LC64		0.720	10.61	0.82	
	LS		−0.343	1.12	−0.08	
	CS		−0.375	1.18	−0.11	
	LI		0.071	1.14	0.06	
	CI		0.119	1.90	0.10	0.92

Table 5.21A: Regression results, 1970 — Liberal-exit contests

Dependent	Independent	a	b	t	beta	R^2
Conservative-held seats						
CS		0.830				
	CL66		0.021	0.34	0.08	
	CI		−0.001	0.05	−0.01	0.01
LS		0.882				
	CL66		−0.173	2.68	−0.54	
	LI		0.092	1.19	0.23	0.24
CL70		0.661				
	CL66		0.788	12.34	0.89	
	CS		0.036	0.17	0.01	
	LS		−0.397	1.94	−0.14	
	CI		0.033	0.57	0.03	
	LI		0.199	3.03	0.18	0.93
Labour-held seats						
LS		0.875				
	LC66		−0.094	3.18	−0.49	
	LI		0.013	0.18	0.03	0.21
CS		0.950				
	LC66		−0.146	4.70	−0.64	
	CI		0.098	1.30	0.21	0.38
LC70		0.697				
	LC66		0.471	12.16	0.87	
	LS		0.066	0.25	0.02	
	CS		−0.365	1.45	−0.15	
	LI		0.022	0.31	0.02	
	CI		−0.001	0.09	−0.00	0.90

Table 5.22A: Regression results, 1959 — Liberal-entry contests, Conservative-held seats

Dependent	Independent	a	b	t	beta	R^2
CS		1.009				
	CL55		-0.022	1.45	-0.16	
	CI		-0.026	1.25	-0.14	0.05
LS		1.182				
	CL55		-0.028	7.28	-0.63	
	LI		-0.032	0.82	-0.07	0.40
CL59		-0.281				
	CL55		1.280	31.62	0.94	
	CS		0.265	1.12	0.03	
	LS		-0.269	2.17	0.07	
	BS		-0.050	0.35	-0.01	
	CI		0.047	1.10	0.03	
	LI		0.068	1.59	0.04	0.96
BS		0.552				
	CL55		0.001	0.08	-0.00	0.01
CB59		4.896				
	CL55		0.453	2.02	0.24	
	CL		-0.969	0.72	-0.07	
	BS		-2.813	5.04	-0.47	
	LS		1.466	2.13	0.25	
	CI		-0.402	1.69	-0.16	0.36
LB59		3.613				
	CL55		-0.671	3.93	-0.38	
	LS		2.014	3.84	0.37	
	BS		-1.921	4.53	-0.35	
	CS		-0.524	0.52	-0.04	
	LI		-0.234	1.30	-0.10	0.58

Table 5.23A: Regression results, 1959 — Liberal-entry contests, Labour-held seats

Dependent	Independent	a	b	t	beta	R^2
LS		1.029				
	LC55		−0.116	3.10	−0.52	
	LI		0.051	0.86	0.16	0.21
CS		1.069				
	LC55		−0.122	4.79	−0.69	
	CI		0.001	0.06	0.00	0.43
LC59		−0.616				
	LC55		1.170	16.98	1.02	
	LS		0.227	0.90	0.04	
	CS		0.202	0.54	0.03	
	BS		−0.101	0.20	−0.02	
	LI		−0.015	0.20	−0.00	
	CI		−0.015	0.022	−0.01	0.95
BS		0.632				
	LC55		−0.095	1.51	−0.29	0.11
LB59		9.358				
	LC55		1.193	0.58	0.16	
	LS		4.781	0.63	0.15	
	BS		−0.284	0.06	−0.01	
	CS		−11.326	0.97	−0.27	
	LI		2.079	0.92	0.19	0.01
CB59		4.736				
	LC55		−0.238	0.24	−0.07	
	CS		−4.196	0.76	−0.22	
	BS		−0.835	0.38	−0.08	
	LS		3.605	1.00	0.24	
	CI		−0.608	0.57	−0.12	0.02

Table 5.24A: Regression results, 1964 — Liberal-entry contests, Conservative-held seats

Dependent	Independent	a	b	t	beta	R^2
CS		1.141				
	CL59		−0.075	5.12	−0.45	
	CI		−0.040	2.48	−0.22	0.20
LS		1.132				
	CL59		−0.275	9.80	−0.69	
	LI		0.001	0.07	0.00	0.47
CL64		−0.546				
	CL59		1.064	23.10	1.03	
	CS		0.034	0.17	0.00	
	LS		−0.272	2.52	−0.10	
	BS		0.018	0.31	0.01	
	CI		0.012	0.35	0.01	
	LI		0.010	0.24	0.01	0.91
BS		0.569				
	CL59		0.039	0.77	0.08	0.01
CB64		2.089				
	CL59		−0.121	0.49	−0.06	
	CS		1.154	1.06	0.10	
	BS		−1.845	5.79	−0.48	
	LS		1.234	2.16	0.25	
	CI		0.169	0.91	0.08	0.30
LB64		3.572				
	CL59		−1.349	5.82	−0.55	
	LS		0.872	1.60	0.14	
	BS		−1.484	4.89	−0.31	
	CS		1.032	1.00	0.07	
	LI		−0.066	0.30	−0.02	0.58

Table 5.25A: Regression results, 1964 — Liberal-entry contests, Labour-held seats

Dependent	Independent	a	b	t	beta	R^2
LS		1.172				
	LC59		-0.162	2.57	-0.38	
	LI		-0.060	0.96	-0.14	0.17
CS		1.083				
	LC59		-0.145	2.40	-0.37	
	CI		-0.042	0.74	-0.11	0.10
LC64		0.171				
	LC59		1.411	17.31	0.91	
	LS		-0.324	1.48	-0.09	
	CS		-0.130	0.53	-0.03	
	BS		0.103	0.47	0.02	
	LI		0.067	0.88	0.04	
	CI		-0.168	2.32	-0.11	0.92
BDS		0.376				
	LC59		-0.039	0.65	-0.10	0.01
LB64		2.898				
	LC59		0.847	2.57	0.41	
	LS		1.499	1.69	0.31	
	BS		-1.237	1.38	-0.23	
	CS		1.608	1.63	0.31	
	LI		0.345	1.11	0.17	0.22
CB64		3.060				
	LC59		-0.927	3.30	-0.50	
	CS		1.389	1.65	0.29	
	BS		-1.366	1.79	-0.28	
	LS		-0.357	0.47	-0.08	
	CI		0.384	1.54	0.22	0.31

Table 5.26A: Regression results, 1970 — Liberal-entry contests, Labour-held seats

Dependent	Independent	a	b	t	beta	R²
LS		0.864				
	LC66		-0.043	1.91	-0.29	
	LI		-0.022	0.53	-0.08	0.15
CS		1.060				
	LC66		-0.155	7.96	-0.78	
	CI		0.035	0.97	0.10	0.62
LC70		0.493				
	LC66		0.823	18.71	0.93	
	LS		-0.303	1.55	-0.05	
	CS		-0.309	1.38	-0.07	
	BS		0.124	0.72	0.02	
	LI		-0.016	0.29	-0.01	
	CI		0.056	1.05	0.04	0.96
BS		0.365				
	LC66		-0.024	0.92	-0.14	0.04
LB70		18.580				
	LC66		0.207	2.16	0.35	
	LS		-5.530	1.01	-0.18	
	BS		-8.305	1.76	-0.29	
	CS		-0.045	0.01	-0.00	
	LI		1.015	0.71	0.12	0.19
CB70		6.495				
	LC66		-1.562	1.92	-0.46	
	CS		-0.791	0.19	-0.05	
	BS		-4.342	1.36	-0.21	
	LS		0.660	0.18	0.02	
	CI		0.923	0.95	0.15	0.14

Table 5.27A: Regression results, 1974 — Liberal-entry contests, Labour-held seats

Dependent	Independent	a	b	t	beta	R^2
LS		0.814				
	LC4F		−0.092	3.50	−0.43	
	LI		0.122	2.16	0.27	0.21
CS		0.780				
	LC4F		−0.149	5.14	−0.59	
	CI		−0.044	0.62	−0.07	0.33
LC40		−0.199				
	LC4F		1.500	19.50	0.93	
	LS		−0.200	0.63	−0.03	
	CS		−0.349	1.20	−0.05	
	BS		0.586	1.39	0.05	
	LI		0.061	0.47	0.02	
	CI		−0.045	0.31	−0.01	0.94
BS		0.303				
	LC4F		−0.025	1.28	−0.18	0.06
LB40		2.027				
	LC4F		0.952	4.95	0.75	
	LS		0.462	0.41	0.08	
	BS		−1.949	1.67	−0.22	
	CS		0.657	0.90	0.13	
	LI		−0.073	0.20	−0.02	0.37
CB40		1.531				
	LC4F		−0.277	3.12	−0.39	
	CS		1.104	3.30	0.39	
	BS		−0.560	1.14	−0.11	
	LS		0.533	1.45	0.16	
	CI		0.046	0.27	0.03	0.56

Table 5.28A: Regression results, 1983 — Conservative-held seats, Liberal for Alliance

Dependent	Independent	a	b	t	beta	R^2
CS		0.851				
	CL79		−0.049	1.93	−0.14	
	CI		0.017	1.71	0.13	0.03
LS		0.718				
	CL79		−0.076	7.48	−0.48	
	LI		0.215	2.42	0.14	0.27
CL83		0.777				
	CL79		1.267	15.92	0.68	
	CS		1.366	1.87	0.07	
	LS		−2.801	5.82	−0.24	
	AS		1.538	3.51	0.13	
	CI		−0.014	0.05	−0.00	
	LI		−0.422	0.67	−0.03	0.78
AS		0.978				
	CA79		−0.121	5.41	−0.57	
	LA79		0.001	0.02	0.00	
	AI		0.319	2.30	0.14	0.34
CA83		1.543				
	CA79		0.203	8.43	0.53	
	CS		0.133	0.80	0.04	
	AS		−0.545	4.82	−0.30	
	LS		−0.250	2.48	−0.13	
	CI		0.072	1.29	0.07	
	AI		−0.478	2.23	−0.11	0.55
LA83		0.062				
	LA79		0.268	14.48	0.66	
	LS		0.366	4.58	0.19	
	AS		−0.157	2.30	−0.08	
	CS		−0.001	0.10	−0.01	
	LI		0.549	6.31	0.20	
	AI		−0.224	1.68	−0.05	0.84

Table 5.29A: Regression results, 1983 — Conservative-held seats, SDP for Alliance

Dependent	Independent	a	b	t	beta	R^2
CS		0.911				
	CL79		−0.017	1.27	−0.11	
	CI		−0.024	0.97	−0.08	0.01
LS		1.074				
	CL79		−0.225	9.68	−0.64	
	LI		−0.004	0.04	−0.00	0.40
CL83		−1.702				
	CL79		2.213	27.66	0.90	
	CS		0.882	2.06	0.06	
	LS		−0.576	2.49	−0.08	
	AS		0.329	1.59	0.04	
	CI		0.041	0.36	0.01	
	LI		−0.443	1.39	−0.04	0.92
AS		0.922				
	CA79		−0.089	2.96	−0.46	
	LA79		0.045	1.58	0.23	
	AI		0.306	3.24	0.26	0.13
CA83		2.185				
	CA79		0.084	3.79	0.30	
	CS		−0.209	0.90	−0.07	
	AS		−0.578	4.98	−0.39	
	LS		−0.002	0.01	−0.00	
	CI		0.091	1.55	1.11	
	AI		−0.356	2.55	0.20	0.34
LA83		0.351				
	LA79		0.181	6.06	0.45	
	LS		0.725	4.74	0.36	
	AS		−0.136	0.95	−0.06	
	CS		−0.358	1.25	−0.08	
	LI		0.595	2.92	0.17	
	AI		−0.144	0.89	−0.16	0.55

147

Table 5.30A: Regression results, 1983 — Labour-held seats, Liberal for Alliance

Dependent	Independent	a	b	t	beta	R^2
LS		0.974				
	LC79		−0.111	3.00	−0.33	
	LI		−0.040	0.77	−0.09	0.09
CS		1.001				
	LC79		−0.244	6.98	−0.63	
	CI		0.074	0.37	0.04	0.39
LC83		−0.579				
	LC79		1.089	13.60	0.93	
	LS		0.337	1.58	0.10	
	CS		−0.004	0.20	−0.01	
	AS		0.111	0.62	0.04	
	LI		0.073	0.81	0.04	
	CI		−0.265	0.76	−0.04	0.78
AS		0.466				
	LA79		−0.046	2.55	−0.36	
	CA79		0.028	1.32	0.18	
	AI		0.638	4.21	0.44	0.23
LA83		0.734				
	CA79		0.282	8.55	0.65	
	LS		0.644	2.57	0.17	
	AS		−0.770	2.48	−0.23	
	CS		0.952	3.17	0.28	
	LI		0.192	1.36	0.11	
	AI		−0.897	2.07	−0.18	0.60
CA83		0.330				
	CA79		0.275	9.96	0.66	
	CS		1.018	5.06	0.38	
	AS		−0.530	2.65	−0.19	
	LS		−0.257	1.28	−0.08	
	CI		0.005	0.01	0.00	
	AI		−0.386	1.50	−0.10	0.73

Table 5.31A: Regression results, 1983 — Labour-held seats, SDP for Alliance

Dependent	Independent	a	b	t	beta	R^2
LS		0.920				
	LC79		−0.099	3.16	−0.29	
	LI		0.013	0.40	0.04	0.07
CS		1.020				
	LC79		−0.258	7.09	−0.61	
	CI		0.035	0.27	0.02	0.36
LC83		−0.002				
	LC79		0.877	17.54	0.89	
	LS		0.189	1.37	0.06	
	CS		−0.114	0.91	−0.05	
	AS		−0.001	0.08	−0.01	
	LI		0.071	1.54	0.07	
	CI		−0.001	0.05	−0.00	0.82
AS		0.476				
	LA79		−0.054	3.17	−0.31	
	CA79		0.112	4.15	0.42	
	AI		0.328	5.62	0.44	0.32
LA		1.209				
	LA79		0.150	3.52	0.30	
	LS		0.333	0.86	0.08	
	AS		−0.813	2.91	−0.29	
	CS		−0.179	0.55	−0.05	
	LI		0.449	3.38	0.29	
	AI		−0.207	0.95	0.09	0.31
CA83		0.753				
	CA79		0.103	2.49	0.22	
	CS		0.998	5.21	0.48	
	AS		−0.785	4.62	−0.45	
	LS		0.344	1.42	0.13	
	CI		−0.233	0.81	−0.07	
	AI		−0.356	3.09	0.28	0.35

6

Nationalist Parties, Minor Parties, Sheffield City Council and European Elections

So far, all of the discussion in this study has been of elections in England and, with regard to those elections, of the three main parties/groupings which have contested general elections therein since 1950 (i.e. Conservative, Labour, and Liberal/Alliance). Scotland, Wales and Northern Ireland were excluded because their separate party systems — an increasingly strong nationalist party in the first two, and an entirely different set of parties in the third — meant that they could not readily be incorporated into the models developed in Chapter 3. Separate investigations were called for. Initial exploration suggested that meaningful results would not be obtained for Northern Ireland because of the small number of constituencies (13 until 1979; 17 in 1983), so no analyses were conducted. (In addition, the strongly polarised party system, based very largely upon religious affiliation, suggested that spending would have little impact, while party fragmentation and realignments within the two religious groupings made derivation of a baseline difficult to establish.) Scotland and Wales also have relatively few constituencies, given the need to subdivide into Conservative-held and Labour-held seats for the models. Nevertheless, it was decided to analyse the 1974 (October) and 1979 elections, to evaluate the impact of spending by the nationalist parties.

The so-called main parties (Conservative; Labour; Liberal/ Alliance) have dominated English elections since the Second World War; no other party has either presented itself as a national force (the closest has been the National Front, which contested 306 seats in England in 1979), and very few individuals have been elected to the House of Commons who were not the official candidates of Conservative, Labour, Liberal or the SDP. Nevertheless, many others do try, so that in 1979 there were 2074 candidates for 516

English constituencies (an average of 4.02 per constituency) and in 1984 there were 2000 for 523 seats (average 3.82). Those 'other' candidates — almost all of whom were either independents (i.e. had no affiliation to a party which was fielding more than one candidate) or nominees of minor parties (e.g. National Front; Ecology) — were subject to the same limitations on campaign expenditure as the others. For most of these candidates, campaign funds (in addition to the deposit of £150, which the great majority forfeited) were hard to obtain, and so sums spent would be small. However, the absence of any link to a major party with access to the media suggests that any money spent on publicity by them should be much more important in providing information than that spent by the 'major party' candidates and could bring substantial benefits in terms of votes won. This hypothesis has been tested for the 1979 general election.

Other elections conducted in Britain are subject to similar limitations on money spent on campaigns. All candidates for election to local councils are subject to spending limits, for example, so the models developed here should be modified for that context. The spending returns are not collected centrally, however, although they are available locally for public scrutiny, at a cost, for a stipulated period after an election. Data were collected and analysed for one local authority — Sheffield City Council — for a period of three elections (1980, 1982, 1983; spending data were available for the past two dates only). No claim can be made that Sheffield is typical of big cities, let alone of all local authorities, but the results are briefly reported here.

Two further elections subject to similar legal constraints have been held recently — those to the European Assembly. For both, England had 66 constituencies which were all contested by Conservative, Labour and Liberal/Alliance. (Scotland and Wales also had nationalist candidates, and were therefore excluded from this study; Northern Ireland elections were not held on a constituency basis.) Compared with general elections, these European elections — in 1979 and 1984 — were not the subject of much public interest. In general, too, the candidates were not well-known to the electorate, and so were unable to stimulate interest by their presence. Thus, the provision of information about them through advertising and other forms of publicity (the average constituency, an amalgam of several House of Commons constituencies, contained over 540,000 electors) should bring electoral benefits.

Each of these four contexts — nationalist parties in Scotland and Wales; minor parties and independents in England in 1979; Sheffield

City Council elections in 1982 and 1983; and the European Assembly elections in England in 1979 and 1984 — provide further situations for testing the hypotheses outlined in Chapter 3 regarding the pattern and impact of campaign spending. However, because of peculiarities in each, the models of Figures 3.2–3.5 cannot be tested since certain elements of them are not relevant to the particular context. Hence, separate models are developed here for each set of elections, as outlined in the following sections.

WELSH AND SCOTTISH NATIONALISTS

During the period under consideration in this study, electoral support for the Scottish and Welsh nationalist parties (the Scottish National Party and Plaid Cymru) has increased substantially. In 1951, together they polled only 18,219 votes, or 0.1 per cent of the British total; by 1959 this total had increased fivefold to 99,309, and it doubled again to 189,545 in 1966. The two elections of 1974 saw the parties win not only their largest number of votes (804,554 in February; 1,005,938 in October), but also substantial Parliamentary representation with 9 MPs (7 for the SNP) in February and 14 (11 for the SNP) in October. (The SNP had won a single seat in 1970.) Their support fell substantially thereafter, to 636,890 votes in 1979 and 457,676 in 1983, with only four MPs (two for each party) returned at each election. Nevertheless, they remain potent electoral forces, and their vote-winning campaigns are worthy of consideration. They are both based on deep political traditions (Agnew, 1984; Cooke, 1984), which provide the foundations for the modern revival; in 1979 Plaid Cymru contested all but two of the 36 Welsh seats (Monmouth and Newport) and the SNP fielded a candidate in every seat in Scotland.

Given their relatively recent appearance as an electoral force, therefore, it is worthwhile considering the degree to which their vote-winning was influenced by the amount which they spent on the campaigns. For both, although their appeal was national in scope, nevertheless, as both Agnew and Cooke show, their ability to win support is at least partly based on their mobilisation of voters in local communities and cultures, a process that should be assisted by spending on publicity during the election campaign.

The hypotheses tested here are the two main ones enunciated earlier — the marginality hypothesis and the party-spending hypothesis. Thus, an extra variable (PCS — Plaid Cymru spending) was added to the basic model (Figure 3.3) for three-party contests in Wales, and a similar variable SNS (SNP spending) to the model for Scotland. The expectations are that the better the chance of the nationalist party overhauling one of its opponents (i.e. the smaller the vote ratio), the more it would spend; and the more that it spent, *ceteris paribus*, the more votes it would get. The hypotheses were tested for the October 1974 and 1979 general elections. Because of the small number of observations, especially in Wales, the validity of some of the results is marginal in terms of statistical significance.

The analyses were initially run only for those constituencies which had both a nationalist and a Liberal candidate at both of the elections in the pair; i.e. they were of four-party contests. (Further analyses produced no substantial differences from the results reported here.) The full results are not given. Instead, Tables 6.1–6.4 present only the regression coefficients relating to the two hypotheses tested: the impact of the three vote ratios at the first election on the level of nationalist spending; and the impact of nationalist spending in the second election on the six vote ratios (CL, CB, LB, CN, LN, BN).

The SNP spent substantial sums of money during the 1974 campaign, on average outspending both Liberal and Labour in the Conservative-held seats, and Liberal in the Labour-held (Table 6.1). There was little relationship between the vote ratios and its spending, however, although in Labour-held seats, the expected negative correlation with the LN ratio at the February 1974 election was obtained. The impact of its spending was nil, in the statistical terms applied here; not one of the twelve regression coefficients even approached significance.

In 1979, the SNP again spent relatively large sums of money (especially so for a minor party), outspending Labour by an average of 12 percentage points and Liberal by 27 points in Conservative-held seats, and outspending Liberal by an average of 28 points in Labour-held seats. Again, with regard to the pattern of its spending, the only clear relationship was with the Labour:SNP ratio; the greater the difference between the two parties in October 1974, the less that the SNP spent in 1979 (Table 6.2). The impact of that spending was much more in line with the hypotheses that was the case in 1974, however. In the Conservative-held seats, SNP spending cut into both the Conservative:SNP and the Labour:SNP ratios; in the Labour-held seats it eroded the latter only. (In Labour-held seats, it also had a negative influence on the Labour:Conservative

Table 6.1: Spending and votes for SNP, 1974 (October) — Four-party contests[a]

Conservative-held seats

Mean SNP Spending 65.3   Standard Deviation 27.2
Regression Coefficients for SNP Spending (with t values) on

CNF −0.093 (0.65)   LNF −0.208 (1.68)   BNF   0.090 (0.69)

Regression Coefficients for Vote Ratios (with t values) on SNP Spending

CLO   0.591 (0.95)   CBO   0.862 (1.07)   CNO −0.020 (0.09)
LBO −0.479 (1.58)   LNO   0.030 (0.13)   BNO −0.051 (0.37)

Labour-held seats

Mean SNP Spending 50.7   Standard Deviation 25.6
Regression Coefficients for SNP Spending (with t values) on

LNF −0.210 (4.84)   CNF −0.145 (0.23)   BNF −0.115 (1.69)

Regression Coefficients for Vote Ratios (with t values) on SNP Spending

LCO −0.676 (0.96)   LBO   1.261 (0.39)   LNO −0.250 (0.24)
CBO   0.285 (0.64)   CNO   0.118 (0.40)   BNO 0.133 (0.29)

Note: a. C, L, B, N relate to Conservative, Labour, Liberal and Nationalist respectively; F refers to the election in February 1974 and O to the election in October 1974. Thus, CN4 is the Conservative:Nationalist ratio in February 1974.

Table 6.2: Spending and votes for SNP, 1979 — Four-party contests[a]

Conservative-held seats

Mean SNP Spending 69.1   Standard Deviation 21.7
Regression Coefficients for SNP Spending (with t values) on

CNO   0.463 (1.40)   LNO −0.381 (1.57)   BNO −0.468 (1.70)

Regression Coefficients for Vote Ratios (with t values) on SNP Spending

CL9 −0.360 (0.72)   CB9 −1.192 (0.29)   CN9 −3.017 (4.09)
LB9   0.346 (0.36)   LN9 −1.543 (2.66)   BN9 −0.590 (1.58)

Labour-held seats

Mean SNP Spending 58.4   Standard Deviation 20.6
Regression Coefficients for SNP Spending (with t values) on

LNO −0.356 (1.94)   CNO   0.167 (1.03)   BNO −0.028 (0.07)

Regression Coefficients for Vote Ratios (with t values) on SNP Spending

LC9 −0.808 (2.02)   LB9   0.495 (0.24)   LN9 −2.667 (2.55)
CB9   1.584 (1.46)   CN9 −0.030 (0.05)   BN9 −0.901 (1.08)

Note: a. For a key to acronyms, see Table 6.1; 9 refers to the general election in 1979.

Table 6.3: Spending and votes for Plaid Cymru, 1974 (October)
— Four-party contests[a]

Conservative-held seats

Mean Plaid Cymru Spending 32.7   Standard Deviation 18.7
Regression Coefficients for Plaid Cymru Spending (with t values) on

CNF   0.0   LNF -0.010 (0.23)   BNF   0.001 (0.08)

Regression Coefficients for Vote Ratios (with t values) on Plaid Cymru
Spending

CLO -0.044 (0.49)   CBO   0.297 (0.38)   CNO -3.573 (0.75)
LBO -0.358 (0.56)   LNO -3.230 (0.75)   BNO -1.439 (0.79)

Labour-held seats

Mean Plaid Cymru Spending 44.7   Standard Deviation 28.2
Regression Coefficients for Plaid Cymru Spending (with t values) on

LNF -0.052 (2.14)   CNF   0.008 (0.28)   BNF   0.083 (1.28)

Regression Coefficients for Vote Ratios (with t values) on Plaid Cymru
Spending

LCO   1.471 (1.84)   LBO -2.853 (0.50)   LNO -12.815 (3.05)
CBO -1.405 (0.43)   CNO -6.357 (3.02)   BNO -4.383 (2.75)

Note: a. For key see Table 6.1.

ratio, suggesting that the SNP campaign activity was more damaging
to Labour — the party which had failed to deliver on its devolution
promises during the previous Parliament — than to Conservative —
the party of the Union with England: presumably, voters were
penalising Labour rather than voting for Conservatives.)

The small number of constituencies in Wales produces great
difficulties in testing the hypotheses statistically, especially in the
Conservative-held seats. Plaid Cymru was clearly not as able to raise
money for its campaign as the SNP; in 1974 it spent less, on
average, than any of the other three parties in Conservative-held
seats, but more than the Liberals (by 10 points) in Labour-held seats.
The pattern of its spending showed no relation at all to the vote ratios
in the eight Conservative-held seats (Table 6.3), but it did spend
more, the closer its February vote was to that of Labour in the 23
Labour-held seats. In the former seats, its spending had no impact.
In the Labour-held seats, however, Plaid Cymru spending had the
expected significant negative impact on all three of its vote ratios:
it appears that the more that it spent on the campaign the more votes
it won from all three parties.

In 1979, Plaid Cymru spent more on average in Conservative-

Table 6.4: Spending and votes for Plaid Cymru, 1979 — Four-party contests[a]

Conservative-held seats					

Mean Plaid Cymru Spending 44.5   Standard Deviation 31.5
Regression Coefficients for Plaid Cymru Spending (with t values) on

CNO	0.0	LNO	−0.001 (0.24)	BNO	0.017 (0.20)

Regression Coefficients for Vote Ratios (with t values) on Plaid Cymru Spending

CL9	0.001 (0.02)	CB9	− 2.138 (0.68)	CN9	0.0
LB9	− 0.991 (0.45)	LN9	− 3.418 (5.60)	BN9	− 10.671 (5.16)

Labour-held seats

Mean Plaid Cymru Spending 35.6   Standard Deviation 28.1
Regression Coefficients for Plaid Cymru Spending (with t values) on

LNO	0.001 (0.01)	CNO	0.042 (0.52)	BNO	− 0.048 (0.44)

Regression Coefficients for Vote Ratios (with t values) on Plaid Cymru Spending

LC9	3.298 (2.41)	LB9	11.084 (3.35)	LN9	− 25.305 (1.48)
CB9	0.260 (0.14)	CN9	− 19.624 (1.61)	BN9	− 6.587 (2.46)

Note: a. For key see Tables 6.1 and 6.2.

than in Labour-held seats (Table 6.4); it remained the lowest spender in the former, however, and the third highest in the latter (outspending the Liberals by on average 13 points). There were no significant relationships whatsoever between the pattern of its spending and the vote ratios at the preceding election. The impact of the spending was also less clear-cut, largely, it seems, because of statistical problems (as betrayed by the large regression coefficients). In Conservative-held seats, Plaid Cymru spending appears to have helped in its contests with Labour and Liberal, but not with the incumbent party; in Labour-held seats, it produced the anticipated effects on its own performance, again with a lesser impact on the incumbent party, and aided Labour, it seems, in its contests with Conservative and Liberal.

In general terms, these analyses provide further substantiating evidence favouring the party spending hypothesis — bearing in mind the statistical problems arising from the small numbers of observations, especially in the analyses of Conservative-held seats. Like the minor parties in England, the nationalist parties in Scotland and Wales have been able to win some electoral support by spending on advertising and publicity during the campaign period. It appears that voters can be wooed via advertising in the particular circumstances of appeals to nationalist feelings.

MINOR PARTY AND INDEPENDENT CANDIDATES

To date, this study has made no reference to the substantial number of minor party and independent candidates, most of whom win relatively few votes. Some of these receive national publicity, and some involve well-known candidates (such as Tariq Ali of the Socialist Unity party, Vanessa Redgrave of the Workers Revolutionary party, and David Sutch of the Monster Raving Loonie party). The majority are unknowns (both politically and more generally), however, and to win any votes (other than perhaps a random few — either mistaken or protests) must project themselves to the electorate. For them, money spent on constituency publicity campaigns should bring dividends — albeit small and electorally insignificant — in the form of votes.

Whether this was the case in 1979 was tested for four small parties contesting English constituencies — the National Front, the Workers Revolutionary party, the Ecology party, and the Communist party — together with two other groups: those calling themselves Independents, and a catch-all 'Other' category. Initial explorations indicated — not surprisingly — that the level of spending (summarised in Table 6.5) was related to no other variables in the models tested in Chapters 4 and 5.

For the impact of spending on the outcome, three tests were run: for Conservative-held seats, for Labour-held seats, and for all seats. There were five independent variables in each, with the minor party's share of the vote (MPV) as the dependent. The vote ratios

Table 6.5: Spending by minor parties and independent candidates, 1979

	Conservative-held Seats			Labour-held Seats			All Seats		
	N	Mean	SD	N	Mean	SD	N	Mean	SD
National Front	146	7.7	5.2	147	8.5	5.8	293	8.1	5.5
Workers Revolutionary	19	18.4	3.6	30	19.1	4.5	58	18.7	4.2
Ecology	21	9.7	7.0	25	11.0	8.7	46	10.1	7.9
Communist	10	19.2	6.9	11	25.6	13.4	21	22.6	11.0
Independents	22	22.1	23.6	27	13.3	19.1	49	18.0	21.4
Others	18	8.1	8.4	23	14.6	13.8	41	11.8	12.1

Figure 6.1: The models tested in the analyses of minor parties and independent candidates. The ratios CL4 etc. refer to the Conservative and Labour vote shares in 1974 (October); A(CL4) means that the variable is the absolute value of the ratio, irrespective of sign

CONSERVATIVE - HELD **LABOUR - HELD**

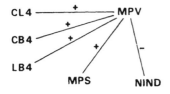   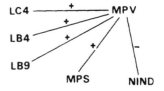

ALL SEATS

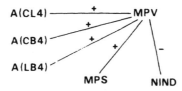

between each pair of the three main parties were included, with the hypothesis that the greater the ratios (i.e. the safer the seat), the greater the potential for vote-winning by a minor party/independent candidate. Spending by the latter (MPS) was expected to have a positive impact on its proportion of the votes cast. Finally, the number of other minor party/independents (NIND) in the constituency was expected to have a negative influence: the greater the number of candidates, the smaller their average share of the vote. (For diagrammatic representations of the models, see Figure 6.1.) In the analysis of all seats, the vote ratios were manipulated as the difference from 1.0; the greater the value, the safer the seat.

On average, these parties and candidates spent but a small proportion of the allowed maximum, with only the Communist party in Labour-held seats and Independents in Conservative-held seats producing means of greater than 20 per cent (Table 6.5). In general terms, it was the two parties of the extreme left of the British political spectrum — Communist and Workers Revolutionary —

Table 6.6: The impact of spending by minor parties and independent candidates, 1979: Conservative-held seats

---

National Front

$MPV = 0.008 + 0.064$ MPS
 (0.011)

$R^2 = 0.21$

Workers Revolutionary

$MPV = 0.016 + 0.087$ MPS $+ 0.004$ CL4
 (0.019) (0.001)

$R^2 = 0.56$

Ecology

  no significant variables

Communist

  no significant variables

Independents

$MPV = -0.011 + 0.224$ MPS
 (0.058)

$R^2 = 0.44$

Others

$MPV = 0.011 + 0.033$ MPS
 (0.012)

$R^2 = 0.55$

---

which spent most; the latter had a very small standard deviation. There were some substantial differences between Conservative- and Labour-held seats, especially among the 'non-party' candidates: Independents spent much more in Conservative-held seats, and Others in Labour-held.

The models were tested in this case using stepwise multiple regression, with a cut-off excluding all variables insignificant at the 5 per cent level. The final equations are given in Tables 6.6–6.8. The average value of R^2 is surprisingly high, at 0.24; for the 14 analyses with at least one significant independent variable, it is 0.30. The two best-fitting equations both refer to Conservative-held seats, where over half of the variation in Workers Revolutionary party and Others' votes was accounted for — in the latter case by the spending variable alone. The poorest performance is for the Communist

159

Table 6.7: The impact of spending by minor parties and independent candidates, 1979: Labour-held seats

National Front

$$MPV = 0.001 + 0.079 \text{ MPS} + 0.003 \text{ LC4} + 0.003 \text{ NIND}$$
$$ (0.014) (0.001) (0.001)$$

$R^2 = 0.30$

Workers Revolutionary

$$MPV = 0.001 + 0.001 \text{ LB4}$$
$$ (0.000)$$

$R^2 = 0.16$

Ecology

$$MPV = 0.017 - 0.007 \text{ NIND} + 0.003 \text{ LB4}$$
$$ (0.002) (0.001)$$

$R^2 = 0.048$

Communist

  no significant variables

Independents

$$MPV = 0.006 + 0.045 \text{ MPS}$$
$$ (0.015)$$

$R^2 = 0.31$

Others

  no significant variables

party; the most consistent — across all three analyses — is for Independents, with R^2 values of 0.44, 0.31, and 0.35, in each case associated with the spending variable alone.

The level of spending was by far the most common significant influence on the electoral performance of the minor parties and independent candidates, being significant in nine of the eighteen equations. It had no impact whatsoever on the vote-winning prospects of the Communist and Ecology parties (perhaps surprisingly in the latter case, since the Ecology party was clearly concerned with issues that could readily be conveyed to the electorate by publicity material), and it only influenced the Workers Revolutionary party support in Conservative-held seats. For the National Front, on the other hand, it was apparently responsible for delivering a substantial number of votes, relative to the party's

Table 6.8: The impact of spending by minor parties and independent candidates, 1970: all seats

---

National Front

$MPV = 0.008 + 0.074$ MPS $+ 0.002$ LB4 $- 0.002$ CB4
$\qquad\qquad\quad (0.009) \qquad\;\; (0.0001) \qquad (0.0001)$

$R^2 = 0.26$

Workers Revolutionary

$MPV = 0.008 + 0.001$ NIND
$\qquad\qquad\qquad (0.0001)$

$R^2 = 0.07$

Ecology

$MPV = 0.019 - 0.004$ NIND $+ 0.001$ LB4
$\qquad\qquad\quad (0.001) \qquad\quad (0.0001)$

$R^2 = 0.22$

Communist

$MPV = 0.025 - 0.003$ NIND
$\qquad\qquad\qquad (0.001)$

$R^2 = 0.11$

Independents

$MPV = 0.004 + 0.158$ MPS
$\qquad\qquad\qquad (0.034)$

$R^2 = 0.35$

Others

$MPV = 0.013 + 0.036$ MPS $- 0.005$ CB4
$\qquad\qquad\quad (0.011) \qquad\quad (0.001)$

$R^2 = 0.24$

---

average performance; in a constituency where it spent the maximum, it could have anticipated 6–8 percentage points more of the poll than in one where it spent nothing (assuming that the trends identified here can be extrapolated over the full range — see p. 188: it averaged only 1.5 per cent of the poll in the seats contested).

It was the Independents and, to a lesser extent in Conservative-held seats only, the Others who benefited most from their campaign spending. These are candidates with no party link which might bring a core of votes in a constituency, and who are therefore dependent entirely on their personal characteristics and presentation. Some are already public figures, contesting the election in order to get more (relatively cheap) publicity — although those are more common at by-elections than at general elections. Some are making a point (often obscure) simply by standing, and make little or no attempt to win votes during the campaign. Others are committed to a particular policy (usually a specific issue) and work hard to project it to the electorate in the hope of winning support. The last group are the most likely to spend resources on the campaign and, as the regression equations indicate, they benefit very substantially from it. In situations of almost complete ignorance prior to the election, therefore, publicity helps to win votes.

Of the other independent variables in the models, there is little evidence of further consistent influences on the performance of minor parties and Independents. The variable LB4 appears in four, suggesting that the weaker the Liberal party relative to Labour, the greater the vote-winning potential of the Workers Revolutionary and Ecology parties and the National Front (the first two in Labour-held seats). NIND appears five times, in all cases but one with the expected negative sign; the greater the number of minor party and Independent candidates in a contest, the smaller the percentage of the vote won by Communist, Ecology and Workers Revolutionary party candidates — though not by Independents and Others.

If one assumes that advertising and other forms of publicity are usually of greatest value — i.e. produce the best returns on investment — in situations of relative ignorance of the 'commodity' being promoted, then in the context of English politics it is readily deduced that minor parties and Independents should get excellent returns from any investment which they are able to make for electoral publicity. The results of the regression analyses reported in this section provide broad empirical support for that deduction: publicity, it seems, can win over some of the 'ignorant'. Furthermore, the returns are quite substantial in relative terms. In Conservative-held seats, for example, the Workers Revolutionary Party could increase its vote more than five-fold if it spent the maximum allowable; and in Labour-held seats, Independents could increase their vote more than seven-fold if they were able to raise about £4500 for the campaign. In contrast, Conservative spending at the same election (1979) would increase the average Conservative:Labour ratio by only abour 40 per cent (see Table 5.2) if the full amount were spent: if Labour candidates spent the full amount challenging in those constituencies they would reduce the average ratio by about 20 per cent only. Minor party and Independent candidates are not disadvantaged by the small amounts of money which most are able to spend to the extent that their chances of winning the seats are diminished; those chances are virtually nil in any case. Nor, in most cases, would most candidates save their deposits if they spent more. However, money does allow them to make a relatively greater impact on the electorate.

SHEFFIELD CITY COUNCIL

Elections to Sheffield City Council are held in three years out of every four — until 1981, elections to the now-defunct South Yorkshire County Council were held in the fourth year. Ostensibly, local government elections are concerned with local issues, and candidates are elected to provide efficient services, with particular reference to the needs of their ward constituents. As such, their publicity campaigns are part of the political evaluation process immediately prior to the election and should be extremely relevant, with a substantial impact. However, in recent decades local politics has become 'nationalised', especially in the large cities, and voting at local elections is as much a passing of opinion by the electorate on the government in office at national level as it is a response to local issues. Variations over time in support for the parties are quite substantial in some places, not because of large-scale conversions during the campaign period but rather because of differential turnout. Given this nationalisation, the local campaign may be largely irrelevant; electors decide whether or not to cast votes on their attitudes to national issues rather than the local ones promoted by the candidates.

The model to be tested here is the same as that for the Three-party contests (Figure 3.3), except that since so few defeated candidates contested the same wards again, only the candidate incumbency variable for the party holding the seat was used. Data were available to allow the models to be tested for the 1982 election, with 1980 as the baseline, and for the 1983 election (with 1982 as the baseline); incumbents were councillors elected four years earlier (one-third of the council is elected each time) and so they had a substantial period of public and local exposure, which some used more than others. Sheffield City Council has traditionally been dominated by the Labour party (Hampton, 1970), and of the 29 wards the great majority are represented by Labour councillors. Thus, it was possible to test the model for Labour-held wards only (Figure 6.2);

Figure 6.2: The models tested in the analysis of Sheffield City Council elections

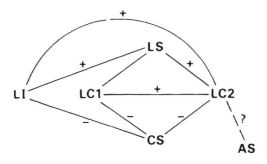

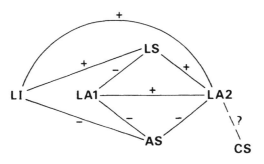

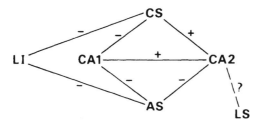

because of their long period of public exposure, and the relative lack of information to the electorate about other candidates, it was suggested that Labour incumbents would not only have a positive influence on the amount of money their party spent, but a negative influence on the amount that challengers would bother to raise in contesting the seats which they held. Not all wards were contested

Table 6.9: Regression results, Sheffield City Council elections, 1982

Dependent	Independent	a	b	t	beta	R^2
LS82		0.185				
	LC80		−0.035	0.85	−0.33	
	LI		0.013	0.06	0.02	0.11
CS82		0.041				
	LC80		−0.027	2.63	−0.67	
	LI		−0.068	1.32	−0.33	0.50
AS82		0.114				
	LA80		−0.009	1.63	−0.57	
	LI		−0.269	1.70	−0.60	0.19
LC82		−1.455				
	LC80		1.579	13.55	0.91	
	LS82		0.996	1.06	0.06	
	CS82		−4.857	2.05	−0.11	
	AS82		2.206	1.87	0.11	
	LI		−1.478	4.11	−0.17	0.99
LA82		3.469				
	LA80		0.043	0.70	0.36	
	LS82		0.578	0.18	0.08	
	AS82		−4.300	0.90	−0.51	
	CS82		−0.986	1.18	−0.49	
	LI		−2.508	1.26	−0.66	0.05
CA82		0.738				
	CA80		0.027	1.12	0.30	
	CS82		0.528	0.44	0.08	
	AS82		−2.253	2.49	−0.81	
	LS82		0.744	1.30	0.33	0.81

Table 6.10: Regression results, Sheffield City Council elections, 1982–Labour:Conservative contests only

Dependent	Independent	a	b	t	beta	R^2
LS82		0.161				
	LC80		−0.031	1.40	−0.31	
	LI		0.086	0.46	0.19	0.10
CS82		0.158				
	LC80		−0.023	4.15	−0.67	
	LI		−0.058	1.19	−0.19	0.45
LC82		0.837				
	LC80		1.299	7.68	0.83	
	LS82		0.709	0.57	0.05	
	CS82		−8.172	1.72	−0.18	
	LI		−2.086	2.06	−0.16	0.89

Table 6.11: Regression results, Sheffield City Council elections, 1983

Dependent	Independent	a	b	t	beta	R²
LS83		0.683				
	LC82		−0.114	1.38	−0.67	
	LI		−0.122	1.28	−0.33	0.01
CS83		0.192				
	LC82		−0.018	1.89	−0.59	
	LI		−0.071	1.91	−0.35	0.48
AS83		1.092				
	LA82		−0.286	2.14	−0.84	
	CA82		−0.005	0.02	−0.01	
	LI		−0.189	1.94	−0.40	0.33
LC83		−0.224				
	LC82		0.792	7.21	0.96	
	LS83		−0.076	0.05	−0.01	
	CS83		−0.155	0.05	−0.01	
	AS83		1.690	1.34	0.15	
	LI		−0.090	0.16	−0.55	0.89
LA83		5.308				
	LA82		1.200	1.06	0.33	
	LS83		2.482	0.71	0.18	
	AS83		−6.872	2.17	−0.64	
	CS83		−9.292	1.22	−0.38	
	LI		−1.248	1.10	−0.25	0.33
CA83		0.822				
	CA82		1.207	3.44	0.69	
	CS83		1.506	1.10	0.22	
	AS83		−0.592	1.09	−0.55	
	LS83		−0.425	0.67	−0.11	0.71

by Liberal/SDP Alliance candidates in 1982, so two analyses were run for that year — one for the contests among all three parties (nine wards only) and one for the Labour:Conservative contests only (26 wards).

The results for 1982 (Tables 6.9 and 6.10) provide very limited support for the models. In the analysis of all three parties (Table 6.9), both Conservative and Alliance spending were negatively related to the 1980 ratios with Labour; the level of Conservative spending influenced the Labour:Conservative ratio, as did the level of Alliance spending on the Conservative:Alliance ratio. The results for the Labour:Conservative contests only (Table 6.10) are very similar.

The analyses for 1983 (18 wards) are similar to those for 1982 in terms of the pattern of spending (Table 6.11): the closer the contest, the greater the level of spending by the challenger.

However, with regard to the election result in 1983, the only significant relationship was between Alliance spending and the Labour:Alliance ratio.

Overall, these results provide little clear evidence to substantiate the models. Sheffield is, of course, dominated by the Labour party — the average LC ratios in Labour-held seats were 3.39 in 1980, 4.91 in 1982 and 3.55 in 1983. There is very little incentive to spend on the campaign in many wards, and indeed the Conservative candidates spent on average only £75 in 1982 and £60 in 1983; in the Labour-held wards, their averages were £38 and £60, compared with Labour averages of £273 and £225 and Alliance averages (in the seats contested) of £164 and £113. The general trends were as hypothesised; no more could probably be expected in such a situation.

THE EUROPEAN ASSEMBLY ELECTIONS

Two elections to the European Assembly have been held to date, those in 1979 and in 1984. Although it was the intention of the European Commission to achieve agreement for a common electoral system (based on some form of proportional representation), this proved to be impossible, largely because of the opposition of the British government. For both elections, therefore, the British MEPs (though not the Northern Irish) were elected according to the first-past-the-post system, using constituencies which were groupings of the Westminster constituencies. Other aspects of British electoral law also applied, including those relating to electoral expenses. In 1979, the allowed maxima were £5000 per constituency, plus 2p for every entry on the electoral roll; five years later they were £8000 per constituency, plus 3.5p for every entry on the electoral roll. This meant that in the context of British elections, potentially relatively large sums could be spent. The average English constituency in 1979 contained some 520,000 voters, giving a spending maximum of £15,488.62; in 1984 it was 540,000 voters and a maximum of £26,985.60. Just over £1 million could have been spent by each party publicising candidates in 1979, therefore, and £1.7 million in 1984.

These elections provide a useful further context for testing the models developed in this study. In particular, they are of interest because of the relatively high level of voter apathy (turnout was only 31 per cent in England in both 1979 and 1984). Thus, an extensive publicity campaign in pre-election period had the potential to inform a substantial proportion of the electorate, most of whom would have known very little about either the candidates or the issues. All parties found it hard to raise money for their campaigns (undoubtedly a reflection of the voters' apathy): in 1979, the average expenditure per candidate was only about 30 per cent of the maximum possible; and in 1984, as the official return of expenses (Home Office, 1984, p. 1) noted, over 80 per cent of the candidates spent less than half of the allowed maximum, with 40 per cent spending less than 20 per cent. Thus, if campaign spending does have an impact, the few candidates able to spend heavily should have benefited considerably.

Two books written on these elections note the low levels of campaigning activity in general and of spending in particular, but provide no detailed analysis of the impact of the money spent. For 1979, for example, Butler and Marquand (1980) record that only 30 per cent of the allowed total was spent, with Conservative candidates generally spending more than their opponents (p. 99): total Conservative expenditure was almost twice that of the Labour candidates and three times that of the Liberals (p. 128), because the Conservatives, unlike the other two parties, were able to add their own money from central funds to that provided by the European Commission. The greater availability of money for the Conservatives should have enabled them to promote their local campaigns more effectively than did their opponents, but Butler and Marquand did not investigate this. Nor did Butler and Jowett (1985) do so five years later, for again the Conservative candidates spent much more, averaging 43 per cent of the maximum compared with 25 per cent by Labour and 23 per cent by the Alliance.

This greater availability of funds for Conservative local campaigns is clear from Table 6.12. Although the party spent more on average in 1979 defending the seats that it notionally held than in those where Labour had the largest percentage of the votes in the 1979 general election, Conservative spending was also greater than Labour spending in the Labour-held seats (by ten percentage points). The Liberal party raised even less money than did Labour; substantially more was raised in Conservative- than in Labour-held seats. In 1984, both Conservative and Labour spent about the same in the seats which they were expected to win (i.e. those that they 'held' according to the 1983 general election results), but the Conservative party was able to raise much more (about £2500 on average) in the seats where it was the challenger than was Labour in comparable situations. Alliance candidates spent on average less than one-

Table 6.12: The level of spending at the European Assembly elections

	Conservative		Labour		Liberal/ Alliance	
	Mean	SD	Mean	SD	Mean	SD
1979						
Conservative-held seats (N = 45)	54.8	11.5	26.8	15.1	19.8	13.9
Labour-held seats (N = 21)	48.4	11.7	38.8	13.1	13.2	9.5
1984						
Conservative-held seats (N = 54)	47.0	11.2	25.6	12.4	24.9	16.2
Labour-held seats (N = 11)	35.1	15.0	47.6	20.9	22.4	16.5

Figure 6.3: The models tested for the European Assembly elections, 1979: Conservative-held seats

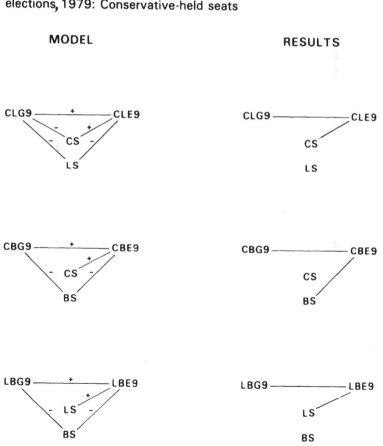

169

Figure 6.4: The models tested for the European Assembly elections, 1979: Labour-held seats

MODEL   RESULTS

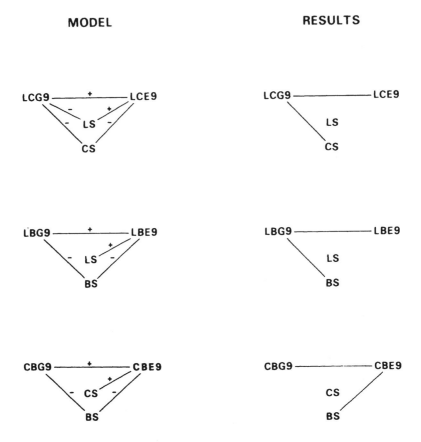

quarter of the maxima available to them.

An evaluation of the pattern and impact of that spending (of not inconsiderable sums — £1.75 million in England in 1984, for example, or 5p per elector) is attempted here, using models similar to those tested in the previous chapters. Four of the hypotheses (p. 48) only are tested for 1979; those excluded relate to candidate incumbency.

For 1979, the models are shown in Figures 6.3 and 6.4. Thus, in Conservative-held seats (defined below), the Conservative: Labour ratio at the European election (CLE9) is positively linked to the ratio (CL69) at the general election in that constituency one month earlier, representing the continuity hypothesis; and similarly, Conservative:Liberal and Labour:Liberal ratios are positively linked. The marginality hypothesis suggests, as before, that the

Figure 6.5: The models tested for the European Assembly elections, 1984: Conservative-held seats

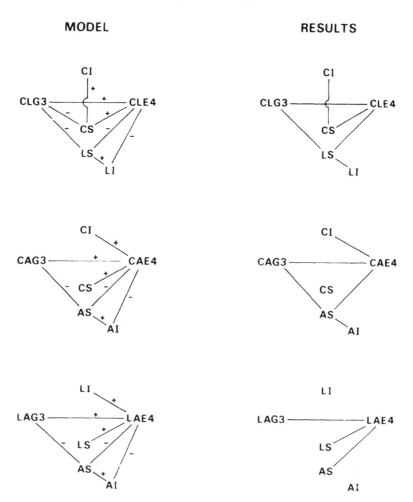

larger the ratio at the first election of the pair, the lower the spending, by all parties; and the party-spending hypothesis suggests that the amount spent significantly affects the outcome of the second contest, in the usual way. The party fund-raising hypothesis is not tested in the regression models.

For 1984 (Figures 6.5 and 6.6), it is possible to test a full model involving all of the hypotheses. In this case, incumbency relates to a candidate who held a European Assembly seat prior to the election. As at the 1983 general election, the Alliance comprised two parties, which divided the seats. The small number of constituencies made separate tests for the two impossible, so a dummy variable (LST)

171

Figure 6.6: The models tested for the European Assembly elections, 1984: Labour-held seats

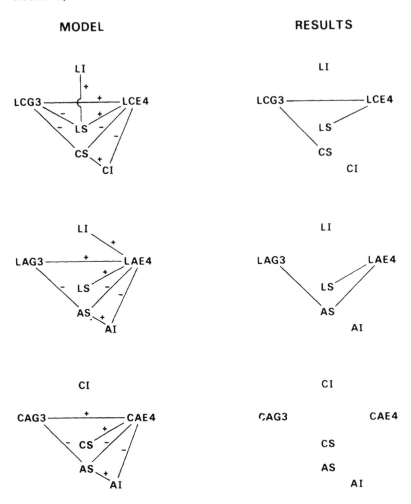

MODEL

RESULTS

was introduced to index a Liberal candidate standing for the Alliance.

In both cases, the implication of the model is that the pattern of voting at the previous general election (one month before in 1979; one year before that of 1984) will be repeated in the European Assembly election. Parties will respond to that pattern in their money-raising and spending, and the amount spent may have some significant influence on the vote ratios at the European contest. The left-hand variables in the models were simply the sums of the votes obtained at the relevant general elections in those amalgams of constituencies.

As already indicated, the Conservative party outspent the others at both elections (Table 6.12), but the overall pattern was consistent with the party fund-raising hypothesis. Continuity in the geography of voting was strong, and only one of the regressions between the two pairs of vote ratios was not significant (for the Conservative: Alliance ratios in Labour-held seats in 1984).

Turning to the marginality hypothesis, only two of the four expected significant relationships were observed in 1979 — for Conservative and Liberal/Alliance challengers in Labour-held seats (Tables 6.13 and 6.14; Figures 6.3 and 6.4). In Conservative-held seats, the probability of a strong Conservative performance immediately after the 1979 general election success was presumably a further deterrent to money-raising for challengers in the seats that Conservative candidates were expected to win (note the regression coefficient of 1.344 for the CLE9:CLG9 link in Table 6.13, indicating a much higher turnout of Conservative than Labour voters in the European Assembly election relative to the general election); Labour and Liberal challengers would have found it hard to raise further money so soon after a general election, and their parties were also electorally demoralised, so the pattern of spending was unlikely to approximate the 'rational' pattern suggested by the marginality hypothesis. In Labour-held seats, on the other hand, Conservative

Table 6.13: Regression results, European Assembly elections, 1979 — Conservative-held seats

Dependent	Independent	a	b	t	beta	R^2
CS		0.523				
	CLG9		0.014	0.38	0.07	0.01
LS		0.289				
	CLG9		−0.045	0.76	−0.17	0.08
CLE9		−0.371				
	CLG9		1.344	15.82	0.91	
	CS		0.711	1.69	0.09	
	LS		−0.272	−0.82	−0.04	
	AS		0.017	0.05	0.01	0.88
BS		0.301				
	CBG9		−0.025	0.40	−0.15	0.01
CBE9		3.411				
	CBG9		0.301	1.84	0.24	
	CS		0.080	0.07	0.01	
	BS		−3.483	3.49	−0.44	
	LS		2.139	2.27	0.29	0.36
LBE9		0.699				
	LBG9		0.683	7.55	0.69	
	LS		1.771	2.56	0.23	
	BS		−1.040	1.45	−0.12	
	CS		−0.432	0.50	−0.04	0.69

Table 6.14: Regression results, European Assembly elections, 1979 — Labour-held seats

Dependent	Independent	a	b	t	beta	R^2
LS		6.484				
	LCG9		−0.087	0.82	−0.19	0.04
CS		0.818				
	LCG9		−0.231	2.41	−0.57	0.21
LCE9		0.209				
	LCG9		0.812	5.31	0.79	
	LS		−0.143	0.49	−0.06	
	CS		−0.312	0.81	−0.12	
	BS		0.367	0.77	0.10	0.70
BS		0.370				
	LBG9		−0.060	2.26	−0.64	0.28
LBE9		3.328				
	LBG9		0.937	3.22	0.59	
	LS		−1.184	0.65	−0.10	
	BS		−4.209	1.36	−0.25	
	CS		−2.837	1.34	−0.20	0.59
CBE9		2.950				
	CBG9		0.833	3.29	0.58	
	CS		−0.351	0.16	−0.02	
	BS		−5.738	2.10	−0.36	
	LS		−0.567	0.30	−0.05	0.52

challengers with a reasonable expectation of success were apparently able to raise more money, presumably in the hope of achieving a differential turnout relative to the general election. (The regression coefficient of 0.812 for LCE9 on LCG9 indicates the relative inability of the Labour party to encourage its supporters to vote again.) Liberal challengers were also able to raise more where the probability of outflanking Labour was greatest, suggesting the use of campaign funds to restore damaged electoral pride.

In 1984, the pattern of spending by incumbent parties was again not significantly related to the marginality of the constituencies (though it was almost so in Labour-held seats), whereas spending by challengers was so related. This is in line with the findings for general elections, and suggests that with a year between the general election of 1983 and the European Assembly election of 1984, the challenging parties were able to assess their electoral chances and raise money accordingly (Tables 6.15 and 6.16). Conservative incumbents presumably felt relatively safe even in marginal seats; Labour incumbents less so.

Regarding the impact of spending, twelve significant links were expected in each year: four were observed in 1979 and eight in 1984

Table 6.15: Regression results, European Assembly elections, 1984 — Conservative-held seats

Dependent	Independent	a	b	t	beta	R²
CS		0.404				
	CLG3		0.009	0.78	0.11	
	CI		0.063	1.82	0.25	
	LST		−0.006	0.21	−0.03	0.02
LS		0.334				
	CLG3		−0.034	3.04	−0.39	
	LI		0.216	1.90	0.24	
	LST		0.010	0.33	0.04	0.20
CLE4		−0.034				
	CLG3		0.593	19.35	0.91	
	CS		0.790	2.21	0.10	
	LS		−0.679	1.79	−0.09	
	AS		−0.131	0.51	−0.02	
	CI		0.079	0.91	0.04	
	LI		0.049	1.68	0.01	
	LST		0.056	0.73	0.10	0.92
AS		0.467				
	CAG3		−0.121	1.82	−0.30	
	LAG3		−0.031	0.59	−0.10	
	AI		0.085	5.43	0.07	
	LST		0.078	1.81	0.24	0.11
CAE4		1.858				
	CAG3		0.177	1.83	0.19	
	CS		0.174	0.50	0.05	
	AS		−1.338	5.47	−0.58	
	LS		0.424	1.42	0.14	
	CI		0.182	2.26	0.22	
	AI		0.967	3.79	0.36	
	LST		0.176	2.51	0.24	0.56
LAE4		0.292				
	LAG3		1.469	12.00	0.80	
	LS		1.214	2.36	0.16	
	AS		−0.653	1.82	−0.11	
	CS		−0.471	0.94	−0.06	
	LI		0.269	0.70	0.04	
	AI		0.185	0.52	0.08	
	LST		0.131	1.28	0.07	0.86

(Figures 6.5 and 6.6). The better performance in the latter year further suggests that the parties were more able at that time to raise funds where those funds were likely to be of most benefit. For 1979, the four significant relationships suggest no general pattern. Liberal spending influenced its share of the vote relative to that of the Conservative party in both Conservative- and Labour-held seats, but had no impact on its contests with Labour. In 1984, the most interesting result is that Labour spending had a significant influence

175

Table 6.16: Regression results, European Assembly elections, 1984 — Labour-held seats

Dependent	Independent	a	b	t	beta	R^2
LS		1.126				
	LCG3		−0.469	1.50	−0.65	
	LI		−0.120	0.72	−0.30	
	LST		0.123	0.70	0.30	0.01
CS		0.952				
	LCG3		−0.489	3.58	−0.88	
	CI		−0.019	0.25	−0.06	
	LST		0.083	1.20	0.28	0.51
LCE4		−2.000				
	LCG3		2.392	3.08	1.08	
	LS		1.811	1.99	0.59	
	CS		−0.761	0.35	−0.18	
	AS		−1.879	0.93	−0.49	
	LI		0.043	0.14	0.04	
	CI		0.252	0.76	0.18	
	LST		−0.142	0.47	−0.11	0.75
AS		0.779				
	LAG3		−0.383	3.17	−0.91	
	CAG3		0.089	0.49	0.13	
	LST		0.128	1.46	0.39	0.43
LAE4		7.557				
	LAG3		−1.446	1.29	−0.42	
	LS		2.001	2.47	0.31	
	AS		−8.015	5.11	−0.97	
	CS		−3.377	1.09	−0.37	
	LI		0.446	1.21	0.17	
	LST		2.081	5.55	0.77	0.94
CAE4		0.016				
	CAG3		0.911	1.40	0.39	
	CS		3.227	1.52	0.84	
	AS		−3.682	1.43	−1.05	
	LS		0.290	0.24	0.10	
	CI		−0.246	0.65	−0.20	
	LST		0.596	2.08	0.52	0.45

in all four contest-types in which the party was involved, suggesting that it was able to use publicity drives successfully in its attempts to erode the Conservative near-hegemony of British representation to the European Assembly after 1979. In addition, Alliance spending was influential in three of the four models, but Conservative spending had a significant impact in only one — on its share in the contest with Labour in Conservative-held seats.

Overall, then, the models provided better fits to the 1984 than the 1979 European Assembly elections, especially in the Conservative-held seats. This is not totally unsurprising because of the contexts of the two elections. Although the date of the 1979 election had been known for some time, and the parties had prepared electoral strategies, there did not exist a great deal of party and public interest — in part because the incumbent government (Labour) was at best lukewarm towards the EEC. The general apathy was then increased by the unexpected general election held only one month before the European Assembly election — as a consequence of defeat for the incumbent government in a House of Commons vote of confidence. Once that general election campaign was over (and the general election was held on the same day as local government elections, which was a further call on public interest, party activists, and party funds), it was almost impossible to revive interest in the European Assembly elections. The Labour and Liberal parties were very short of money, had little time to raise funds, and could whip up little enthusiasm among their members. In such circumstances, the Conservative party was able to capitalist subsequently, winning the majority of the seats — and indeed, able to capitalise on its spending in the seats which it held.

In terms of the level of public interest, as indicated by the turnout, the 1984 European Assembly election differed very little from that held five years earlier. Nevertheless, it occurred in a more 'normal' context, a year after the general election, and was used by the opposition parties as an attempt to demonstrate the Conservative government's unpopularity: it was also the first major election since Neil Kinnock's election as leader of the Labour party, and he was determined to score a personal triumph to establish himself. Thus, fund-raising and spending was much more focused on the electoral potentials, especially in the Conservative-held seats, with the anticipated impacts.

IN SUMMARY

This brief chapter has extended the major analyses of English general elections presented in Chapters 4 and 5 by evaluating the models of Chapter 3 (modified as necessary) in four other British electoral contexts. The results do not provide overwhelming evidence in support of the hypotheses embedded in those models, but they do support many of the conclusions reached earlier in this study.

Of the four contexts studied here, the Sheffield City Council elections and the nationalist parties in Scotland and Wales provided least evidence in supporting the models, in part because of the statistical problems encountered with small numbers of observations — especially so in Conservative-held seats in Wales. Nevertheless, expenditure by both Plaid Cymru and the SNP was apparently an influence on their relative vote-winning capacity, even though the pattern of their spending was unrelated to the local electoral context. For minor parties and Independents in England in 1979, again there was clear evidence that in the case of some candidates the amount spent had a significant and, in relative terms, substantial impact on their electoral fortunes. Finally, for the European Assembly elections, the models proved to be better representations of the situation in 1984 than in that of the relatively unusual context of 1979.

Overall, then, it is reasonable to conclude that these further tests provide additional support for the conclusions drawn from the earlier chapters. For the electoral contexts discussed here, it could have been suggested that campaign expenditure would be more important than for the 'major parties' at English general elections; electoral apathy and unknown candidates were the norm in the contests analysed here. However, as the discussions throughout this book have stressed, money is not necessarily a significant influence on electoral outcome. It depends on the particular context in which it was spent, a conclusion amply reinforced here.

7

Evaluation

This book has presented the first major study of the *impact* of political spending in Britain, with reference to constituency-level campaign expenditure at nine general elections since 1950. It has shown that, contrary to the expressed opinions of other researchers — opinions that have not been based on detailed statistical research — the amount spent on those campaigns is linked to the electoral outcome, in ways entirely consistent with models of the process of political evaluation which draw their analogies from the advertising industry: the greater the expenditure on information provision by a producer, *ceteris paribus*, the greater the market share. The evidence presented here is not overwhelming, since it does not indicate that the level of spending is a major, let alone a dominant, influence on the result. Nevertheless, the results of the analyses cannot readily be gainsaid; in most aspects of British elections since 1950, and especially since 1969 when the limits on spending were raised, the results of the constituency contests have been influenced by the campaign spending of the parties involved.

This final chapter evaluates those results, in two ways. First, it summarises the results of the detailed analyses through a review of the nine hypotheses set out in Chapter 3. Secondly, it explores the implications of those results with regard to the operations of the British electoral system.

THE HYPOTHESES EVALUATED

The model tested in this book is set firmly in the context of that used in the majority of psephological studies in Britain, as characterised by the major outputs of the British Electoral Study. It comprises two major components. The first — encapsulated in the processes of political socialisation — sees voters developing the personal ideologies from which stem their attitudes towards the various political parties; such development involves the local milieu providing interpretations of the national political context. Those attitudes are not permanent, but tend to be consistent over substantial periods of time, so that political socialisation leads most voters to identify with a particular party. The result is long-term continuity, both in individual voting behaviour and in the pattern displayed by population aggregates — people in places.

The second component of the model relates to the decision-making that takes place at the time of each election. It involves the processes of political evaluation, whereby people assess the parties — their policies, candidates, and leaders — in terms of desirability as future providers of national government. Such decision-making may be so habitual that it is brief and shallow; the processes of political socialisation have resulted in such a close alignment between voter and one party that no alternative is even closely scrutinised. However, for many voters — and an increasing number according to proponents of the dealignment thesis — the options are carefully evaluated in the light of available information. The outcome of such evaluation may well be a shift in preference — the party voted for is not the one supported at the previous election — with consequences for the electoral geography and, of course, the electoral outcome.

It is the second of these sets of processes which has been the focus of attention here, since the concern has been the impact of spatial variations in the provision of information by parties to the electors on the pattern of votes. Thus, most of the hypotheses are concerned with the changes between elections, as brought about by campaign spending. They are in two groups, one dealing with the geography of spending (i.e. the inputs) and the other with the geography of the outcome (i.e. the throughputs); a further group considers changes in those geographies over time. Firstly, however, the continuity of voting that should result from political socialisation is evaluated.

The continuity hypothesis

This hypothesis was represented in the models (e.g. Figure 3.2) by a single link that postulated a close positive relationship between the vote ratios at each pair of elections. In the contests between

Conservative and Labour in Conservative-held seats, for example, the C:L ratio in 1950 (where C:L is the ratio between the Conservative and Labour percentage of the votes in each constituency) should be positively related to the C:L ratio for 1951. Moreover, given the general view about the continuity of voting in Britain, that relationship should be by far the strongest in the analysis of the results of the second election: strength was measured by the relative size of the standardised partial regression coefficient (beta).

For most of the period under discussion here, British politics has been dominated by two parties (Conservative and Labour) and in the great majority of constituencies they have been the only realistic contenders for electoral success. (This statement is less true of Scotland and Wales; it is irrelevant to Northern Ireland. Most of the detailed analyses here have been concerned with England.) In addition, the beta coefficients show that over the entire period, the ratio of the vote share between the two parties has been consistent, moving up and down by about the same amount in all constituencies according to the relative national popularity of the two parties. (This is the uniform swing phenomenon noted by Butler and Stokes, 1974, p. 121, and referred to in all of the Nuffield studies.) The electoral geography of the country has thus been remarkably consistent and, as shown in Chapter 5, the importance of other influences on the Conservative:Labour share of the vote has been slight relative to the continuity represented by this hypothesis.

The electoral geography of the Liberal Party has been much less consistent than that of the other two major parties (see Johnston, 1983a): in part because it has not always fielded candidates (especially in the 1950s and 1960s); in part because it has traditionally been the recipient of protest votes against the other two parties, especially in safe seats; and in part because increasingly, Liberal mobilisation of the electorate has been built upon 'grass root politics' which vary substantially in their intensity and interest. Thus, the continuity hypothesis, though generally reflecting the major influence on the pattern of votes at any election, was not as strongly supported by analyses of the Conservative:Liberal and Labour: Liberal shares of the vote as by those of the Conservative: Labour (p. 84). The former ratios were spatially more variable over time, and were more strongly influenced by the pattern of spending (see below).

The political socialisation component of the model is thus clearly substantiated by the analyses here, especially with regard to the vote shares of Conservative and Labour. Continuity is the dominant aspect of the voting behaviour and electoral geography of England.

The pattern of spending

Three hypotheses were incorporated into the model in order to account for variations between constituencies in the amount spent by each party: the *party fund-raising hypothesis* suggested that parties would raise and spend more in seats which they held and were defending, than in seats that were held by other parties (H_2); the *candidate fund-raising hypothesis* suggested that incumbent candidates (either the constituency MP or a loser at the last election who continued to 'nurse' the constituency) would raise more than non-incumbents (H_3); and the *marginality hypothesis* suggested that all parties (both incumbent and challengers) would raise and spend more, the more marginal the constituency was for them (H_4). Of these three hypotheses, there was strong supporting evidence favouring H_2 and H_4, but very little for H_3.

For much of the period, the Conservative party has been better able to raise campaign funds than its Labour challenger, in all constituencies, whereas the Liberal party has generally been able to raise much less (little more than half of the allowed maxima), especially in the 1960s and 1970s. The major differences among the parties emerged after 1969 when the spending maxima were substantially increased.

At almost all of the elections studied, local Conservative parties in constituencies that were held by their MPs raised significantly more to defend their incumbency than did those in constituencies where Conservative candidates were challenging Labour possession, in line with the party fund-raising hypothesis. For the Labour party, this was also the case in a majority of the elections (i.e. significantly more spent in Labour-held than in Conservative-held seats), but there was a number of cases (in all types of contest) which showed no significant difference in Labour spending between the two types of seat. The reason for this, it seems, is linked to the marginality hypothesis: Labour tends to have more very safe seats than Conservative, in which the pressures to spend are not great, especially as many of them are in mining areas where the political socialisation component is extremely strong and political evaluation consequently weak.

At most elections, local Liberal parties raised more in the Conservative-held than in the Labour-held seats, suggesting either that local organisations were stronger and fund-raising easier in the former than in the latter, or that the electoral prospects were better in the Conservative-held seats (or both). Again, this is in part linked to the marginality hypothesis, as suggested above. Furthermore, Liberal candidates (and in 1983, SDP candidates also) have generally performed better in southern England, which contains mostly Conservative-held seats, than in the Labour strongholds further north, and this may be reflected in their spending.

Within the parties, there is little evidence of significant differences between *incumbent and non-incumbent candidates* in fund-raising for either Conservative or Labour. The implication is that it was the strength of the local organisation and its hold on the seat which influenced spending, and that on average the nature of the individual candidate played virtually no role in influencing that situation. For the Liberal party, incumbents did raise more funds, especially during the Liberal revival from 1970 onwards, this presumably being indicative of successful constituency 'nursing'. Non-incumbent Liberals contesting seats which had Liberal candidates at the previous election were in general better able to raise money than their contemporaries in Liberal-entry contests, suggesting the importance of a continuing party organisation in the former.

In addition to explorations of intra-party differences in relation to candidate incumbency, tests were conducted into potential significant differences between parties relating to whether their opponents were fielding incumbents or not. To all intents there were none.

The marginality hypothesis suggests that the closer-run the contest in a constituency, as measured by the vote share ratio at the previous election, the greater the incentive to spend, since influencing a few more voters via the political evaluation processes could prove electorally decisive. The results of the regression analyses show this to have been a valid expectation in the majority of contests: the greater the likelihood of winning or losing a seat, the greater the expenditure on wooing the voters. (For most Liberal candidates, the chances of winning the seat were remote, and yet they also consistently spent more, the smaller the ratio between their performance and that of their opponents, especially opponents who held the seats. This suggests that the relative number of votes achieved in the first election was a good indicator both of the strength of the local party organisation and of the willingness of people to contribute to attempts to build on that foundation: relative success breeds the search for more of the same.)

The three hypotheses reviewed in this section assume rationality in the fund-raising activities of parties: they will work harder where they have something to defend (especially if their hold on it is under threat), and where they have a chance of electoral gain. Of course, as indicated in Chapter 1, it is not the three central (national) parties that are being rational according to the expectations, since they make relatively small contributions to the costs of the constituency campaigns (though the marginal seats do quite frequently attract above-average grants). Thus, it is not three actors whose rationality is being studied, but as many as 1500 (the number of constituencies multiplied by the number of parties), so that the verification of two of the hypotheses (H_2 and H_4) is indicative of widespread appreciation of the demands for campaign funds. In general, the parties react in the ways suggested by the party fund-raising and marginality hypotheses. Whether their candidates have contested the seats before is (with the partial exception of the Liberals) largely irrelevant; it is parties which are being sold in British constituency campaigns, not candidates.

Spending and votes

The two hypotheses in this group are central to the whole concern of this study, since they postulate links between candidacy, spending and vote-winning. The first — the *candidate incumbency hypothesis* (H_5) — suggests that incumbent candidates, both MPs who currently represent the constituency and candidates defeated at the last election who have remained to fight there again, will perform better than non-incumbents: following the American experience, candidates are expected to attract a personal following additional to that which a new candidate for the seat would obtain. (Curtice and Steed (1980) suggested that incumbent MPs gained some electoral advantage over non-incumbents but found no similar widespread evidence of such personal effects in 1983 — Curtice and Steed, 1984: Cain (1983) suggests that the incumbency effect is linked to the level of constituency service provided by an MP.) The sixth hypothesis — the *party-spending hypothesis* — postulates that, *ceteris paribus*, the more that a party spends on the campaign in a constituency, the more votes it obtains.

There is very little evidence to support the candidate incumbency hypothesis. Undoubtedly individual MPs and challengers do perform better that non-incumbents in particular constituencies as a consequence of their 'nursing' activities, and others obtain a personal vote for a variety of reasons not necessarily connected to their political activities. As a general tendency, however, there is no incumbency effect — further evidence to support the argument that British voters evaluate the parties (and their leaders), and not their local candidates.

Regarding the party-spending hypothesis, the evidence is not unequivocal, but the weight of it clearly favours the argument that money spent on advertising parties brings electoral benefits. Four particular points can be stressed in the context of this general conclusion.

(1) First, spending by challengers is more likely to influence the vote outcome than is spending by incumbents since, as already indicated in the discussion of H_2, incumbent parties tend to spend more than their challengers and vary less in the proportion of the maximum spent across the constituencies; with a relative absence of variation, it is not surprising that there is also an absence of correlation.

(2) Of the two main parties, spending by Conservatives when they are incumbents is more likely to bring extra votes than is comparable spending by Labour parties. The reason, it seems, is that more Labour-held seats are very safe.

(3) Of the challengers, it has been the spending by the Liberal party and the minor parties/independents which has had most impact — in part because of greater spatial variability, but also because such spending is more likely to counter voter ignorance and increase awareness of those parties than is spending by the better-known Conservative and Labour parties.

(4) The impact of spending increased over the period, being particularly apparent in the 1983 general election. This finding relates to the final group of hypotheses, discussed next.

Overall, then, it cannot be claimed that the analyses reported here have produced overwhelming evidence of the electoral importance of campaign spending in Britain during the period 1951–84. On the other hand, enough clear evidence in line with expectations has been produced, especially for the Three-party contests and for elections since 1970, to ensure that spending cannot be dismissed as irrelevant to the study of the outcome of British general election campaigns. The implications of this are discussed below.

Changes over time

In absolute terms, the amount of money spent by the parties on constituency campaigns increased during the period. In 1945, for example, the average expenditure per candidate was: Conservative, £780; Labour £595; and Liberal, £532. In 1983, the averages were about five times greater — Conservative, £3320; Labour, £2927; and Alliance, £2525. However, during that same period, the cost-of-living index has increased more than tenfold, so that in real terms, expenditure has fallen substantially. In constant 1984 £s, the respective 1945 and 1983 figures per candidate are: Conservative £10,900 and £3500; Labour £8300 and £3000; and Alliance £7400 and £2500. Thus, expenditure per candidate in real terms is now only about one-third of what it was at the beginning of the period studied here. (All of these figures are taken from Pinto-Duschinsky 1981a, 1985.) Moreover, of course, the electorate has increased in size, from 34.6 million in 1951 (for the UK) to 42.2 million in 1983. The average constituency in 1951 had about 55,400 voters, so that expenditure per voter was about 10p by the Conservative party, compared with 8p and 6p by Labour and Liberal respectively. (Pinto-Duschinsky and others sometimes cite figures for spending per vote cast. Spending per votes available, including those who eventually decide not to vote, appears to be a better measure.) In 1983, the respective figures were, for an average constituency of 65,000, 5p, 4.5p, and 4p.

Given this decline in spending on local campaigns, which has generally been countered by increased spending centrally (Pinto-Duschinsky, 1985), together with the increased importance (because of widespread television ownership) of the subsidy-in-kind provided by free broadcasting (Pinto-Duschinsky, 1981a), it was expected that the importance of such expenditure to the parties would decline between 1951 and 1983, along with its impact. Three hypotheses formalised this expectation, with regard to the links between incumbency and spending (H_7), marginality and spending (H_8), and spending and vote share (H_9).

All three of the hypotheses were falsified by the results of the analyses. Over the decades since the Second World War, spending by the parties has declined, but constituency incumbency and marginality have become more important influences on the 'where' of spending, and the amount spent has apparently become more influential on the election results. In particular, these increases have occurred after the decision to increase the spending maximum in 1969. Before that date, the relationships outlined in the models were generally either weak or non-existent, but since then they have become more apparent. Until 1969, Conservative and Labour parties were able to spend close to the maximum in a large number

186

of constituencies, including those in which they were the challengers. There was relatively little variation, and relatively little impact. The Liberals were less able to raise money, however, and there was greater variation, and a greater impact. After 1969, the average level of spending relative to the maximum decreased and variability increased, especially for challengers, in all three parties. This greater variability was linked to a greater electoral impact: parties could spend more, and where they then did so, they benefited from that expenditure.

AND THE IMPACT?

The results just summarised could be dismissed on a variety of grounds. One is that they are statistical artifacts alone, with the statistically significant differences and relationships having no causal significance, or at least not the causal significance attached to them in this study. Those differences and relationships are incorporated in a model derived by *a priori* reasoning, however, and can only be dismissed if the model itself is found wanting — by the production of a better model which can account for the differences and relationships in a separate way. Until that is done, they are presented here as being substantially as well as statistically real.

A second and more interesting dismissal would be not that the differences and relationships are unreal, but they are of little relevance: after all, the continuity effect dominates British voting behaviour, and all studies of the political evaluation process suggest little or no importance for the information-providing role of local campaigning (the vast majority never mention it, of course). So how relevant are the findings set out here? An earlier exercise (Johnston, 1986a) explored the answer to this by simulating the 1983 election result with different spending levels, though using a slightly different model than that employed here. That exploration is extended in the present section.

One problem of any simulation is that the data on which it is based refer to a particular domain only — the range of observations for the variables — and extrapolation beyond that domain is risky. (Thus, the error terms for forecasts are greater than those for estimates; the latter refer to the observed data points only, whereas the former extend beyond the domain of observed values: Johnston and Wrigley, 1988.) The problems which this raises in the present situation are illustrated in Figure 7.1. In that figure, two partial

Figure 7.1: The problems of simulation when extrapolating beyond the domain of observations. The two lines show the partial regression slopes of the CB ratio on CS and BS. The solid lines refer to those slopes within the domain of the observed values

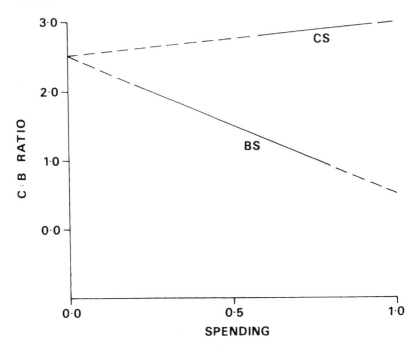

regression slopes for CB on CS and BS are shown: the solid lines refer to the domains of the observed values, so that CS ranged from 0.6 of the allowed maximum to 0.93, whereas BS ranged from 0.20 to 0.80, and the dashed lines extrapolate those relationships to the values of 0.0 and 1.0 for spending.

The slopes in Figure 7.1 suggest the following points:

(1) if neither party spent anything on the campaign, the ratio between their share of the votes would be 2.5;

(2) if Conservative spending remained at zero, the ratio would fall by approximately 0.2 for every 0.1 increase in Liberal spending, so that if BS = 1.0, the ratio is 0.5; and

(3) if Liberal spending remained at zero, the ratio would increase by approximately 0.05 for every 0.1 increase in Conservative spending, so that if CS = 1.0, the ratio is 3.0.

The full regression equation is

$$CB = 2.5 + 0.50CS - 2.00BS$$

188

so that if both spent the maximum

$$CB = 2.5 + 0.50 - 2.00$$

which would put the parties level.

Given that parties spend differentially, a valuable question to ask is 'what would have been the result if they had both spent the same?'; the same could be the level of the highest spender, the maximum, nothing, or some other figure. Using the above equation, we come to the following conclusions for a constituency where $CB = 2.0$, $CS = 0.90$ and $BS = 0.55$.

To simulate different results we define

$$CSD = x - CS \qquad (7.1)$$
$$BSD = x - BS \qquad (7.2)$$

where x is the defined level of spending, and the simulation equation is

$$CB = 2.0 + 0.50CSD - 2.00BSD$$

(a) If both parties spend at the Conservative level, $x = 0.90$ so

$$CB = 2.0 + 0.50(0.0) - 2.00(0.35)$$
$$= 2.0 - 0.70 = 1.30$$

(b) If both parties spend at the Liberal level, $x = 0.55$ so

$$CB = 2.0 + 0.50(-0.35) - 2.00(0.0)$$
$$= 2.0 - 0.175 = 1.825$$

(c) If both parties spend at the maximum, $x = 1.00$ so

$$CB = 2.0 + 0.50(0.10) - 2.00(0.45)$$
$$= 2.0 + 0.05 - 0.90 = 1.15$$

(d) If both parties spend nothing, $x = 0.00$ so

$$CB = 2.0 + 0.50(-0.90) - 2.00(-0.55)$$
$$= 2.0 - 0.45 + 1.10 = 2.65$$

All such simulations are risky if they are used to 'predict' possible outcomes, since they assume not only that the regression slopes are good estimates for the domains (i.e. the equations have high R^2 values and also high partial r^2 values for each variable), but also that the slopes can be extrapolated linearly beyond the specified domains. The further one extrapolates, the greater the standard error of the forecast and the less one's certainty about the prediction. For the present task, such sophisticated forecasting procedures are not used since the goal is only to provide a very general impression of the likely impact of differences in the levels of spending. Only one simulation model is used, enquiring what the result would be in the Three-party contests if all three parties spent to the maximum allowed. This seemed the best simulation strategy, since the result at the maximum is presumably the best that any party can hope for, if it is unaware of how much its opponents will spend.

The steps involved in the simulation were as follows:
(1) for each party, compute the proportion unspent—

$$CSU = 1.0 - CS \qquad (7.3)$$
$$LSU = 1.0 - LS \qquad (7.4)$$
$$BSU = 1.0 - BS \qquad (7.5)$$

(2) for the Conservative:Labour ratio in Conservative-held seats
(a) add to the CL value at the second election (CL2) the value of the unspent Conservative proportion multiplied by the partial regression coefficient for CS(bCS), and
(b) subtract from the CL value (CL2), as modified in (a), the value of the unspent Labour proportion multiplied by the partial regression coefficeint for LS(bLS), giving the new estimate of CL2 (CLE). Thus

$$CLE = CL2 + (bCS)(CSU) - (bLS)(LSU) \qquad (7.6)$$

then
(c) add to CLE the value of the unspent proportion multiplied by the partial regression coefficient for Liberal spending (bBS), giving a further estimate of CL2 (CLEE)

$$CLEE = CLE + (bBS)(BSU) \qquad (7.7)$$

(3) Subtract CLE from CL2 to produce CLC, the change in the value of CL if both Conservative and Labour spent to the allowed maximum

$$CLC = CL2 - CLE \qquad (7.8)$$

(4) subtract CLEE from CL2 to produce CLCC, the change in the value of CL if all three parties spent to the allowed maximum

$$CLCC = CL2 - CLEE \qquad (7.9)$$

(5) repeat steps (3) and (4) for Conservative:Liberal (CCB) and Labour:Liberal (LB) ratios.
Thus, if in a constituency

$$CS = 0.80 \quad LS = 0.75 \quad BS = 0.60$$

then

$$CSU = 0.20 \quad LSU = 0.25 \quad BSU = 0.40$$

If the regression equation for the CL2 ratio includes the following terms (the continuity term is omitted since it is not used in the simulation)

$$CL2 = 1.5 + 0.1CS - 0.2LS + 0.15BS$$

then

$$CLE = 1.5 + (0.1)(0.2) - (0.2)(0.25) = 1.47$$

and

$$CLEE = 1.47 + (0.15)(0.60) = 1.56$$

so that

$$CLC = 1.50 - 1.47 = 0.03$$

and

$$CLCC = 1.50 - 1.56 = -0.06$$

Thus looking at only Conservative and Labour spending, if each spent to the maximum there would be a slight shift to Labour (i.e. CLE is positive); looking at spending by all three parties, the

Conservative lead over Labour would be increased if all spent to the maximum (i.e. CLCC is negative).

These calculations have been undertaken separately for Conservative-held and Labour-held seats with Three-party contests at each of the elections, using the regression coefficients reported in Chapter 5. In each case, they were calculated for the Conservative: Liberal and Labour:Liberal ratios, as well as for the Conservative: Labour (Conservative-held seats) and Labour:Conservative (Labour-held seats). For 1983, there was a subdivision of both Conservative- and Labour-held into those contested by Liberal and those contested by SDP candidates. All of the regression coefficients were used, irrespective of their level of statistical significance, except that an insignificant coefficient with the unexpected sign was replaced by one of 0.001. The resulting new variables are thus illustrative estimates only of the possible impact on the result if each party had spent to the limit.

Tables 7.1 and 7.2 present the average values of the different variables for Conservative- and Labour-held seats with Three-party contests. They show that, in general, the consequence of raising spending to the maximum was to reduce the average ratio, since the majority of the differences (CLC and CLCC in Table 7.1; LCC and LCCC in Table 7.2) are positive. In other words, the challenger would benefit most if each party spent to the maximum allowed — not surprisingly, since in general, challengers spent less than

Table 7.1: Average values of CL and simulated values of CL for Conservative-held seats, Three-party contests

	1951	1959	1964	1966	1970	1974	1979	1983 L	1983 SDP
CL									
Mean	1.10	2.00	1.71	1.51	1.94	1.72	2.49	4.61	2.93
SD	0.14	0.74	0.74	0.62	0.58	0.69	1.18	2.70	1.75
CLE									
Mean	0.94	1.97	1.66	1.54	1.84	1.64	2.44	3.55	2.83
SD	0.15	0.72	0.69	0.62	0.54	0.65	1.15	2.34	1.66
CLEE									
Mean	0.98	2.02	1.74	1.64	1.92	2.05	2.91	4.18	2.96
SD	0.15	0.71	0.68	0.61	0.54	0.46	1.07	2.21	1.67
CLC									
Mean	0.16	0.08	0.05	-0.03	0.10	0.08	0.04	1.15	0.08
SD	0.05	0.05	0.09	0.04	0.08	0.07	0.09	0.48	0.16
CLCC									
Mean	0.12	-0.02	-0.03	-0.12	0.01	0.05	-0.44	0.48	-0.04
SD	0.05	0.07	0.22	0.07	0.08	0.07	0.23	0.62	0.18

Table 7.2: Average values of LC and simulated values of LC for Labour-held seats, Three-party contests

	1951	1959	1964	1966	1970	1974	1979	1983 L	1983 SDP
LC									
Mean	0.95	—	1.85	1.86	1.24	2.03	1.50	1.62	1.33
SD	0.12	—	1.11	1.03	0.47	1.02	0.66	0.72	0.53
LCE									
Mean	1.02	—	1.78	1.82	1.17	1.94	1.42	1.71	1.33
SD	0.12	—	1.02	0.99	0.40	0.98	0.62	0.74	0.53
LCEE									
Mean	—	—	1.70	1.92	1.15	2.09	1.21	1.78	1.33
SD	—	—	1.01	1.00	0.40	0.99	0.61	0.74	0.53
LCC									
Mean	-0.07	—	0.07	0.04	0.07	0.08	0.09	-0.09	-0.03
SD	0.05	—	0.10	0.09	0.09	0.07	0.08	0.08	0.05
LCCC									
Mean	—	—	0.14	-0.07	0.09	-0.06	0.29	-0.16	-0.3
SD	—	—	0.13	0.09	0.09	0.08	0.10	0.08	0.05

Table 7.3: The shift in seats if spending changed: Three-party contests

	1951	1959	1964	1966	1970	1974	1979	1983 L	1983 SDP
Conservative-held seats									
CL < 1.0									
No Change in Spending	8	1	14	9	0	5	0	0	1
C and L Spend Maximum	105	1	9	9	0	8	0	8	2
All Spend Maximum	87	0	4	2	0	3	0	0	1
Labour-held seats									
LC < 1.0									
No Change in Spending	23	—	0	1	15	0	36	11	27
L and C Spend Maximum	12	—	0	2	20	0	46	10	28
All Spend Maximum	—	—	0	1	22	0	95	5	28

incumbents and had steeper partial slopes for the party-spending hypothesis. The amount of change varies between elections, reflecting changes in the relative level of spending. (Note that the results of simulating the values of LCEE in 1951 were nonsensical — i.e. negative ratios — so these were omitted from Table 7.2.)

What would have been the impact of these shifts in terms of seats won? Table 7.3 shows the number of constituencies with CL ratios less than 1.0 in Conservative-held seats and LC ratios less than 1.0 in Labour-held seats, and compares these with the numbers after the simulations. The number of seats 'at risk' was very large at a few elections (notably in 1951 for Conservative-held seats, when the incumbent party received no benefit at all from its spending — Table 5.1A). With the exception of 1974, the greatest potential alteration in the balance of power from different spending levels was in Labour-held seats after 1966. In 1970 and 1979, Labour could have lost substantially more seats to Conservative if the latter had spent to the limits (see Table 4.2 for the amounts spent).

Turning to the ratios between the vote shares of the incumbent parties and those of Liberal challengers, Tables 7.4 and 7.5 generally indicate much more substantial shifts than was the case in Tables 7.1 and 7.2, in all but two cases positive and thus in favour of the Liberal party (the Alliance parties in 1983). This is not surprising, since — as Chapter 4 made very clear — the Liberals have spent substantially less than the other parties throughout the

Table 7.4: Average values of CB and simulated values of CB for Conservative-held seats, Three-party contests

	1951	1959	1964	1966	1970	1974	1979	1983 L	1983 SDP
CB									
Mean	1.26	3.20	2.58	3.12	4.05	2.15	3.42	1.92	2.02
SD	0.28	0.98	0.81	1.02	1.27	0.59	1.12	0.43	0.33
CBE									
Mean	—	2.74	2.24	2.80	2.99	1.83	2.58	1.71	1.86
SD	—	0.79	0.67	0.95	1.12	0.50	0.93	0.36	0.30
CBEE									
Mean	—	2.85	2.38	3.13	3.46	2.05	2.78	1.60	1.86
SD	—	0.78	0.64	0.87	1.06	0.46	0.90	0.32	0.29
CBC									
Mean	—	0.45	0.33	0.32	1.06	0.32	0.83	0.21	0.03
SD	—	0.28	0.30	0.18	0.35	0.19	0.35	0.13	0.04
CBCC									
Mean	—	0.35	0.20	-0.01	0.58	0.10	0.63	0.32	0.03
SD	—	0.30	0.35	0.32	0.41	0.25	0.38	0.14	0.04

Table 7.5: Average values of LB and simulated values of LB for Labour-held seats, Three-party contests

	1951	1959	1964	1966	1970	1974	1979	1983 L	1983 SDP
LB									
Mean	1.17	–	3.34	5.07	5.15	3.60	5.42	2.29	1.98
SD	0.31	–	1.01	2.83	1.87	1.47	1.73	0.79	0.76
LBE									
Mean	0.83	–	3.18	3.75	3.03	2.50	2.06	1.93	1.71
SD	0.46	–	0.96	2.74	1.47	1.36	1.51	0.75	0.71
LBEE									
Mean	1.20	–	3.18	3.36	2.99	1.89	2.00	2.31	1.64
SD	0.44	–	0.96	2.72	1.46	1.23	1.49	0.78	0.71
LBC									
Mean	0.34	–	0.16	1.32	2.12	1.29	3.32	0.36	0.26
SD	0.26	–	0.10	0.60	0.93	0.36	0.64	0.22	0.21
LBCC									
Mean	-0.02	–	0.16	1.71	2.16	1.91	3.39	-0.02	0.33
SD	0.23	–	0.10	0.77	0.94	0.69	0.65	0.30	0.23

period, so their anticipated gains from more expenditure would be greater. Even so, on average they would still have trailed far behind the other two parties in the allocation of votes, with few ratios less than 2.0 prior to 1983. Thus, as Table 7.6 indicates, few seats were 'at risk', certainly until the 1970s when the Liberals spent very little in the Labour-held seats (less than 33 per cent on average) and so could have reaped substantial electoral gains from a tripling of their expenditure. Nevertheless, as the bottom panel of Table 7.6 indicates, greater spending would have made many more seats marginal (if a ratio of less than 1.4 is taken as indicating marginality).

It must be stressed once again that these simulations are purely indicative: *they do not show what the result would have been, only what it might have been.* Averages have been used as if every constituency were like the average, but this may not have been the case; in a marginal seat the Liberals may have got much less benefit from extra spending than in a safe seat, for example. Moreover, as indicated in Chapter 3, most of the spending distributions are heteroscedastic, so that some of the simulations will have been strongly influenced by the patterns in very small numbers of constituencies. What is reasonable for description is not necessarily valid for forecasting. Nevertheless, what has been presented in this section does show that spending is not irrelevant, and that the

Table 7.6: The shift in seats if spending changed: Three-party contests

	1951	1959	1964	1966	1970	1974	1979	1983 L	1983 SDP
Conservative-held seats: C:B Ratios									
C:B < 1.0									
No Change in Spending	0	0	2	1	0	0	0	1	0
C and B Both Spend Maximum	–	0	2	1	2	4	2	5	0
All Parties Spend Maximum	–	0	1	0	0	0	1	6	0
C:B ≤ 1.4									
No Change in Spending	123	3	10	6	1	29	12	26	5
C and B Both Spend Maximum	–	4	17	7	6	58	23	41	9
All Parties Spend Maxmum	–	3	5	3	2	18	12	64	9
Labour-held seats: L:B Ratios									
L:B < 1.0									
No Change in Spending	0	–	0	2	0	0	1	3	7
L and B Both Spend Maximum	39	–	0	1	0	2	52	6	8
All Parties Spend Maximum	16	–	0	2	0	28	55	1	10
L:B ≤ 1.4									
No Change in Spending	49	–	0	2	2	0	1	6	20
L and B Both Spend Maximum	51	–	0	2	2	30	87	21	37
All parties Spend Maximum	45	–	0	3	2	77	86	13	49

statistical significance of the findings reported here can be linked to a possible substantial outcome. However, we do not know what the consequence would be if all parties spent to the maximum in all seats until they do so. (And that would create major analytical problems, because there would be no variation in the spending data! The results would have to be compared with what might have been if the spending pattern had been the same as at the previous election.)

THE POLITICS OF PARTY FINANCE

The issue of party finance and its political implications is one which is frequently raised in the United Kingdom, in the context of beliefs that 'Elections should not only be fairly conducted, but they should be thought to be fairly conducted' (Rose, 1961, p.1). This is a canon, far from universally shared among politicians and commentators alike, that the differential access to campaign funds makes for an inequitable democratic system which favours the party most able to attract substantial sponsorship — the Conservative party. For some commentators, as was clear in the debates over constituency maxima up to 1949, ceilings on electoral spending are the solution. For others, such ceilings are also potentially undemocratic since they restrict the flow of information; the solution to the problem, they claim, is state subsidy for political parties.

The analyses in this study relate to only one aspect of party spending — by constituency parties during election campaigns. They cannot provide material to contribute to the wider debate. Nevertheless, they do throw light on an issue that is rarely, if ever, addressed in contributions to the debate: what is the electoral impact of campaign spending? A Canadian writer has tackled this directly by asking 'Can you spend your way into the [Canadian] House of Commons?' (Isenberg, 1980), concluding that '. . . A candidate's best chance of being elected was to spend more money than anyone else in his or her riding . . .', and 'The most *effective* way of spending money appeared to be on printed messages such as signs and newspaper advertisements' (p. 58). The analysis on which these conclusions were based was fairly superficial, but if they also apply in the UK, they have substantial implications for the political debate.

Many of the issues involved in the debate over British political finance lie outside the scope of this book. Rose (1961) identifies five, for example: (1) the controls on spending relate only to the candidates in the individual constituencies and not to the parties nationally (except, of course, with regard to control of access to the broadcasting media); (2) the controls do not ensure equality of expenditure by the parties; (3) spending between election campaigns, when most people change their minds, is not controlled; (4) expenditure by pressure groups outside the parties is not controlled; (5) 'Although corruption by bribery is outlawed, no controls are placed upon the alleged corruption of voters' free choice by covert public relations' (p. 2; note also O'Leary's (1962) reference to the argument that 'large-scale promises of social

197

benefits . . . were a subtle counterpart to the more open bribery and treating of the past . . . individual bribery was succeeded by bulk purchase' — p. 232). None of these five issues is directly related to the aspect of spending studied here; nevertheless, the results analysed and interpreted are of considerable relevance to the general issue.

The major problem with any research into the impact of spending is that correlation and cause are too readily inflated. Rose (1961, p. 11) notes this when he argues that

> Most advocates of stricter accounting of political expenditure assume that money brings votes; some charge that it buys votes in sufficient quantities to win elections. This assertion is truest when it is most platitudinous: a party cannot operate without money. To go further, and say that a party such as the Liberals gains few votes because it has little money is to mistake cause and effect. It would be more nearly true to say that a party with relatively few voters, such as the Liberals, has difficulty in raising money. As the rise of the Labour Party shows, the necessary minimum is not great, nor is it impossible to secure if the party has strong support in the electorate.

This is very much in line with Butler's views, cited earlier (p. 34 ff); parties spend where they are strong and where seats are marginal, but because all do so, that spending therefore has no impact. In fact, all do *not* do so, to the same extent, as the analyses reported here indicate (recall the statement in Chapter 3, p. 58, that the correlation between spending by two parties is usually relatively weak). Unfortunately, such statements are not based upon substantive research.

Rose appears to recognise this point, but largely in a negative way. He reports that

> Many British discussions of political expenditure seem to assume a simple input-output model of electioneering: X thousand pounds will provide X or X/2 or X/4 or 2X or 4X votes. Y inches of advertising space will produce Y/2 or 2Y units of political influence . . . People unaccustomed to dealing with large sums of money might think it incredible that hundreds of thousands of pounds might be spent to no real effect (p. 11).

By innuendo he falsifies the hypothesis of the first two sentences, without any research to back him up. Like Butler and Kavanagh (1984), he would argue that 'In 1983, therefore, candidates went through the familiar ritual of canvassing, leafleting, addressing public meetings, holding local press conferences, and participating in debates' (p. 245), and conclude that the ritual is nothing more; it does not influence the result. He states that elections are determined by three factors — the material and social environment, individual values, and party activities — and he does not undertake a very full analysis of spending as part of the third factor (his approach compounds all spending — p. 13).

The lack of a research base on which to evaluate the impact of political spending remains today, 25 years after Rose's paper. (Note Bogdanor's (1984) statement that 'the increasing volatility of voting behaviour has made the campaign period more important as a time when votes can be won than it was, say, in the 1950s', which he immediately qualifies by stating that 'it has yet to be proved that campaign expenditures [he was not referring specifically to constituency spending] by a party can improve its prospects in the absence of a potentially well-disposed electorate': p. 140.) In 1975, the then Lord President of the Council (Michael Foot) established a committee under Lord Houghton with the following terms of reference:

> To consider whether, in the interests of Parliamentary democracy, provision should be made from public funds to assist political parties in carrying out their functions outside Parliament; to examine the practice of other Parliamentary democracies, and to make recommendations as to the scope of political activities to which any such provision should relate and the method of its allocation (Houghton Report, 1976, p. iv).

A majority of the Committee recommended in favour of subsidies, in two forms

> (1) annual grants to be paid from Exchequer funds to the central organisations of the parties for their general purposes, the amounts being determined according to the extent of each party's electoral support
> (2) at local level, a limited reimbursement of the election expenses of Parliamentary and local government candidates (p. xv).

With regard to the latter

> Reimbursement should be restricted to those candidates who poll
> at least one-eighth of the votes cast, and the amount to be reim-
> bursed should be the candidate's actual election expenses up to a
> limit of half his legally permitted maximum expenditure.
> Payment should be made directly to the candidate (p. xv).

In other words, only the stronger candidates (those not losing their
deposits by winning at least 12.5 per cent of the votes cast) were
eligible, and then only for half of their expenditure. If money is of
value in vote-winning, those who needed it most were to be denied
it, and all would still need to raise half of the maximum; the evidence
of recent years suggests that this would be much easier for some than
others.

The case for this recommendation regarding a subsidy to
candidates was based very largely on a survey of one hundred constituency parties, which covered such issues as membership and
finance. The report notes that

> Constituency parties were asked if they thought their party was
> experiencing serious, moderate, minor, or no financial difficulty.
> In reply most organisations said that they were in serious or
> moderate financial difficulties; one in ten reported minor
> difficulties, and few said they had no financial difficulties (p. 38).

This led them to conclude that the basis of parliamentary democracy
was under threat, since their chapter in 'The case for state aid'
begins with the statements that

> Effective political parties are the crux of democratic government.
> Without them democracy withers and decays . . . Their function
> is to maximise the participation of the people in decision-making
> at all levels of government. In short they are the mainspring of
> all the processes of democracy. If parties fail, whether from lack
> of resources or vision, democracy itself will fail (p. 53).

British parties lack finance, and state aid is needed as 'the best, and
perhaps the only, way of arresting the run-down of the parties, and
of starting the process by which their effectiveness can be raised to
an adequate level' (p. 56).

The details of the recommendations rest on an unwillingness to

200

encourage 'frivolous candidatures' (p. 68), so that only those who poll at least one-eighth of the votes cast would qualify. Amongst those, 'we would not want to stimulate unnecessary additional expenditure or remove the need for candidates or constituency parties to raise money themselves' (p. 68). The outcome would clearly be of benefit to the candidates of established political parties, and, of them, to those best able to raise money themselves. At the present time, this is clearly the Conservative party, which already is able to raise close to the maximum in the seats that it holds and much more than the other parties in the seats that it does not hold (Tables 4.2 and 4.3, for example). It could qualify fairly easily for reimbursement in nearly every seat, and would be able to spend very close to the maximum as a consequence. For Labour, the Alliance parties, the minor parties and independents, and the Nationalist parties, the potential benefits would be uncertain. The proposal was that they be *reimbursed* their costs, up to half of the maximum allowed, so that they would first have to raise the money — perhaps by a loan; and reimbursement was conditional on saving the deposit (an additional item of initial expenditure), so that a potential recipient who gambled on being reimbursed by spending half of the maximum and then not qualifying could face a bill of more than £2500!

A report commissioned by the Hansard Society (1981; see also Bogdanor, 1982) made very similar recommendations. It argued that

> Unless the parties have enough money to carry out their activities, democracy cannot function efficiently; nor can democracy work fairly if the sources of party finance lead to over-representation of some interests and the under-representation of others (p. 12).

Consequently, it was argued that popular participation in politics should be encouraged by providing state aid which matched the donations received by a party, as long as it met the same conditions as those suggested in the Houghton Report (p. 37). The state aid would go to the parties centrally; the constituency parties would benefit from the donations, and perhaps also by increased central grants.

Because, in general, challengers are less able to raise money than incumbents, the former would be the major beneficiaries from such schemes. At the present time, with fixed limits to constituency campaign spending, this would clearly be to the advantage of the

Alliance parties, as the analyses in the preceding section indicate. There would be a price, however, since average spending by the Liberal party in Three-party contests has been just under 50 per cent over the whole period, and until recent years many of its candidates have lost their deposits. The benefits would also be greater for Labour than for Conservative; as Chapter 4 indicates, the ability of Labour to raise money to defend the seats it holds has declined much more substantially than the comparable ability of the Conservative party (15 per cent from its peak in 1964 to the 1983 figure, compared with 11 per cent for Conservative), and Labour's ability to challenge in seats that it does not hold has declined even further. It also has increasing problems not only in raising what would be matching funds under the Houghton proposals, but also in saving deposits: 119 were lost in 1983.

The simulations presented above are of course hedged around by the caveats introduced there, so any 'prediction' of the number of seats that could have changed hands if the Houghton recommendations had been made law would be very rash. The general pattern is clear, however. It relates to one issue that the Houghton Committee did not address — how high should the maximum be? None of the evidence presented here can assess the likely impact of raising the maxima; it is likely, as Figure 7.2 suggests, that the slope for C:L on CS would become steeper, and less likely that the slope for C:L on LS would become shallower. It is also likely that the Conservative party would be better able to raise more funds, and remain relatively close to the maxima, than would Labour. Thus, without subsidies, the Conservatives would benefit most from a raising of the limits. With subsidies, its advantage would be less, but the difficulties that the other parties would have in raising even half of the new maxima would make problems for them in realising the potential benefits.

If the maxima were lowered, then the advantages of local Conservative parties in fund-raising would be reduced. The situation would be more akin to the 1950s and early 1960s, when the advantages of more money were few because most local parties (especially Conservative and Labour) were able to raise substantial percentages of the maxima. Overall, however, Labour and the Alliance would be the major beneficiaries, it seems.

The only other major study of British political finance is that of Pinto-Duschinsky (1981a), who provides a very substantial descriptive analysis of spending since the first reforms of the 1830s. He sums up his analysis by concluding 'that the Conservatives have a

Figure 7.2: The possible impact of Conservative (CS) and Labour (LS) spending on the C:L ratio if the spending maximum were to be raised

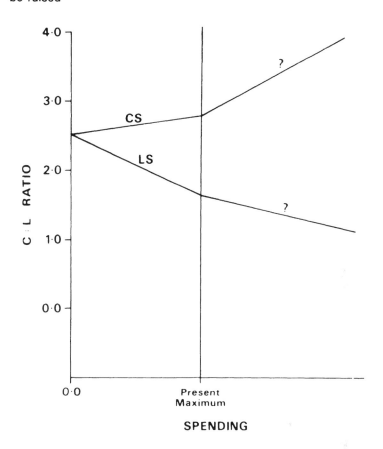

financial advantage over Labour in all aspects of party activity. The Conservative lead is smallest in constituency-level election campaigns. It is greatest at the level of routine constituency finance' (pp. 280–1). The Conservatives, insofar as spending is a reasonable index of activity, are much better able than Labour to nurse constituencies between elections; Labour parties must rely more on the impact of what they do in the campaign period, when they spend relatively more — but still less than the Conservatives in absolute terms. The Liberals are much poorer.

Pinto-Duschinsky is not in favour of state subsidies. Protagonists for such subsidies, especially those who are pro-Labour, argue that the present system is unfair because the Conservative party gets support from 'the backing of big business or from large donations from wealthy individuals' (p. 285). However, he says,

> In fact, the Conservative financial advantage over Labour derives
> almost entirely from fund raising in the constituencies, and this
> in turn results from the fact that local Conservative associations
> have several times more members than their Labour counterparts
> . . . The revenue per member received by the Conservatives is
> small — less than Labour's. It is the relatively wide extent of the
> party's grass-roots support that ensured its financial superiority
> in the 1960s and 1970s. As far as the balance between Conser-
> vative and Labour is concerned, the existing system is not unfair
> (p. 285).

From this, he continues by arguing that

> the introduction of state payments to parties cannot be justified on
> the grounds that they would help to redress existing unfairness in
> the finances of the parties. Indeed, it could be argued that state
> aid would in itself be unfair if it denied the party with the largest
> membership the fruits of its strength at the grass roots (p. 286).

In this, he aligns himself with the dissenting minority of members
of the Houghton Committee, who opposed state aid on the grounds
that political parties are voluntary organisations whose need to raise
funds helped them to maintain links with the electorate. Moreover,
they could find no evidence that subsidies would aid the improve-
ment of party performance: however, like Pinto-Duschinsky, they
did not seek to evaluate the impact of the money currently spent.

Pinto-Duschinsky's analysis and conclusions are based on a very
substantial historical survey which incorporates the following points
(see his summary, p. 292 ff). (1) The limitation of campaign
expenditures from 1883 onwards forced the development of parties
run and operated — especially at election times — by volunteers; this
constituted a major democratising of British politics. (2) When large
personal contributions by candidates were barred after 1945, the
Conservative party (the major beneficiary from such contributions)
had to become much more active locally, which was another healthy
democratisation of politics. (3) Only in the cities, where businesses
continued to provide funds, were active associations not formed, and
these have now atrophied. (4) Until the 1920s, the large central
funds held by the Liberal party — gained, for example, from the sale
of titles — had a retarding effect on the development of local parties,

with probable electoral consequences. (5) Local Labour parties for long relied upon trade union contributions and failed to attract large memberships, and they are now suffering from this. Thus, where local activity is substantial, finances are relatively healthy, and this is best for democracy. (Bogdanor (1984) also argues that the relative financial health of the Conservative party reflects not greater levels of donations by members, but simply an average constituency party five times larger than the average constituency Labour party.)

To the extent that local activity (in terms of numbers of active members who play roles in the processes of political mobilisation and socialisation) and campaign fund-raising are correlated, then the analyses in this study parallel Pinto-Duschinsky's arguments. The more active parties have raised more, and have benefited more accordingly, in terms of electoral returns. If democracy is equated with mass participation, then the more democratic local parties are benefiting. All of the results presented here support that view — parties which spend more generally obtain more votes. However, Pinto-Duschinsky, like Houghton, does not ask how much they should be required to spend. The higher the maxima, the greater the pressure to spend, because of the possible consequences of not so doing. If the parties do differ in their abilities to raise money — and certainly they do differ in the average incomes of their supporters — then the higher the maxima, the greater the potential spending differentials, with the electoral consequences noted here.

The argument would appear to be that spending maxima should be kept low, thereby preventing the development of inter-party differentials. Most local parties should then be able to spend close to the maximum, with the result that spending would have little effect on the electoral outcome (i.e. a return to the 1950s). However, this could be interpreted as undemocratic, in two ways. Firstly, it might remove some of the incentive for local activity, because money is not needed; local parties may atrophy to some extent. Secondly, it could result in less information being provided about the campaign to the electorate, which would lead to a greater centralisation of politics and reliance on the mass media. To avoid these problems, some extension of the subsidies-in-kind might be introduced, to aid in the local presentation of cases.

At the present time, discussion of political finance is somewhat dwarfed by debates over reform of other aspects of the electoral system — notably, the case for proportional representation. The various reforms are not independent, however, since any move to larger constituencies would immediately require greater production of information about candidates for the electorate (as the European Assembly elections indicate). The issues discussed here are bound to remain on the political agenda, therefore. They need to be debated with better analyses of the present situation than heretofore. The present study has attempted to provide such analyses for one aspect of the debate: the use of those analyses is constrained by the contexts to which they refer, but as indications of the links between spending and votes they are far superior to the groundless generalisations that characterise most discussions of the topic.

IN CONCLUSION

This book has come up with neither unequivocal findings nor clear prescriptions. What it has done is expose a set of unsubstantiated general statements about British campaign spending to rigorous statistical testing, and derive some appreciation of how spending has influenced vote-winning in recent decades. No claims of generality for these findings are made: they relate only to the elections studied, and the situation at future elections may be very different.

For any empirical analyses, a theoretical framework is crucial, and such a framework has been provided here by using the model of the various phases of the British Election Study as a basis. This comprises two components — the long-term processes of political socialisation, and the short-term processes of political evaluation, which are separate but interdependent. Within each, people are influenced by the content of their milieux. With regard to political evaluation, this involves the activity of the political parties providing information to influence voter decision-making with regard to current issues and personalities, both national and local. It was argued that, to the extent that such information-providing activity is spatially variable, so the amount of influence on the political evaluation process should similarly vary.

Many aspects of local party activity could be studied within this context, both the continuous activity of inter-election periods and the much greater volume of activity during the few weeks prior to an election. Most of this is difficult to study in a large-scale comparative format because it is not readily quantifiable (though note the Houghton Committee's (1976) research finding that party membership, finance and employment of an agent are all correlated; in general, Conservative local parties are largest, spend most, and are most likely to have a paid agent in their safe seats and least likely in the safe Labour seats, whereas all three indicators suggest greatest Labour activity in the marginal seats which Labour holds — see p. 89 above). One set of data is available that allows detailed quantitative comparisons to be made. This refers to the returns of campaign spending made by the individual candidates under the Representation of the People Act. The precision of the data is somewhat dubious (see p. 33), but there can be little doubt that they do provide a useful general guide to the amount of campaigning activity by each candidate. In particular, because most of the money is reported as spent on publicity materials, they indicate the volume of information provided to the electorate by those seeking their votes.

These data have been used to test a simple model, which states that candidates will raise more funds to protect something which they already hold (a Parliamentary seat) than they will if they are challenging for it; that they will raise more, the greater the stake (i.e. the more marginal the seat); and that the more they spend, the greater the return. It is a simple model: the greater the advertising expenditure, the greater the market share, where campaign information is equated with advertising expenditure, and proportion of the votes cast with market share. The results, summarised earlier in this chapter, are generally supportive of the model, particularly at the elections since 1969 when the maxima that can be spent have frequently been increased.

The interpretation of these results is fairly straightforward. To the extent that the political parties differ in either or both their ability and willingness to raise and make available money for local constituency campaigns, so their relative electoral performance also differs: those who spend more win more where there are differentials (hence the importance of the change in the legislation governing the level of spending). The implications are not so straightforward, as an earlier section of this chapter has indicated. In general terms, it seems desirable that the electorate should be made as fully aware of the options available to them as possible, which suggests equality of spending by the parties. The level of that spending is unclear, however: if it is held down, the advantages to the richer parties are few, but the electorate may be ill-informed; if it is high, the difficulties of the poorer parties are magnified, which leads into the debate on state subsidies.

In their book on the 1983 general election, David Butler and Dennis Kavanagh (1984) recognised the spatial variability in spending but concluded that 'this was not so much a case of money buying victory as of money going to places where victory was probable or possible' (p. 266). This study has suggested that their analysis, which was not based on any detailed statistical manipulation of the data, is over-simple. Certainly, what has been presented here does show that money tends to be raised and spent where 'victory was probable or possible'; but the amounts raised and spent by the various competitors varied, and this was crucial. In many contests, one party could gain a vote advantage over its competitors by spending more. It is difficult to estimate to what extent this altered the election result in terms of the allocation of votes. It certainly influenced the distribution of votes in the constituencies, if the model in which the analyses are embedded is valid. The spending of money in the constituencies on informing and wooing the electorate has won votes in recent (especially the most recent) decades. Whether this raises problems for the operation of a democracy because some would-be MPs are better able to raise money than others remains a moot point.

Bibliography

Agnew, J.A. (1984) Place and political behaviour: the geography of Scottish nationalism. *Political Geography Quarterly, 3,* 191–206

BBC/ITN (1983) *The BBC/ITN guide to the new parliamentary constituencies.* Parliamentary Research Services, Chichester

Berrington, H. (1984) Decade of dealignment. *Political Studies, 32,* 117–20

Berrington, H. (1985) MPs and their constituents in Britain: the history of the relationship. In V. Bogdanor (ed.), *Representatives of the people?,* Gower Press, Aldershot, pp. 15–43

Bochel, J.M. and Denver, D.T. (1971) Canvassing, turnout and party support: an experiment. *British Journal of Political Science, 1,* 257–69

Bochel, J.M. and Denver, D.T. (1972) The impact of the campaign on the results of local government elections. *British Journal of Political Science, 2,* 239–44

Bogdanor, V. (1982) Reflections on British political finance. *Parliamentary Affairs, 35,* 367–80

Bogdanor, V. (1983) *Multi-party politics and the constitution.* Cambridge University Press, Cambridge

Bogdanor, V. (1984) Financing political parties in Britain. In V. Bogdanor (ed.), *Parties and democracy in Britain and America,* Praeger, New York, pp. 127–52

Bogdanor, V. (1985) Introduction. In V. Bogdanor (ed.), *Representatives of the people?,* Gower Press, Aldershot, pp. 1–14

Butler, D.E. (1963) *The electoral system in Britain since 1918.* Clarendon Press, Oxford

Butler, D. and Jowett, P. (1985) *Party strategies in Britain: a study of the 1984 European Elections.* Macmillan, London

Butler, D. and Kavanagh, D. (1974) *The British general election of February 1974.* Macmillan, London

Butler, D. and Kavanagh, D. (1980) *The British general election of 1979.* Macmillan, London

Butler, D. and Kavanagh, D. (1984) *The British general election of 1983.* Macmillan, London

Butler, D. and Marquand, D. (1980) *European elections and British politics.* Longman, London

Butler, D. and Pinto-Duschinsky, M. (1971) *The British general election of 1970.* Macmillan, London

Butler, D. and Stokes, D. (1969) *Political change in Britain.* Penguin, London

Butler, D. and Stokes, D. (1974) *Political change in Britain* (second edition). Macmillan, London

Cain, B.E. (1983) Blessed be the tie that unbinds: constituency work and the vote swing in Great Britain. *Political Studies, 31,* 103–11

Caldeira, G.A. and Patterson, S.C. (1982a) Contextual influences in participation in U.S. State legislature elections. *Legislative Studies Quarterly, 7,* 359–81

Caldeira, G.A. and Patterson, S.C. (1982b) Bringing home the votes: electoral outcomes in State legislature races. *Political Behaviour, 4*, 33–67

Chapman, R.G. and Palda, K.S. (1981) Voting participation in a public consumption perspective. In K.B. Monroe (ed.), *Advances in consumer research VIII*, Association for Consumer Research, Ann Arbor, pp. 530–3

Chapman, R.G. and Palda, K.S. (1983) Electoral turnout in rational voting and consumption perspectives. *Journal of Consumer Research, 9*, 337–46

Chapman, R.G. and Palda, K.S. (1984) Assessing the influence of campaign expenditures on voting behaviour with a comprehensive electoral market model. *Marketing Science, 3*, 207–26

Converse, P.E. (1966) The concept of a normal vote. In A. Campbell, P.E. Converse, W.E. Miller and D.E. Stokes, *Elections and the political order*, John Wiley, New York, pp. 9–39

Cooke, P.N. (1984) Recent theories of political regionalism: a critique and an alternative. *International Journal of Urban and Regional Research, 8*, 549–72

Copeland, G.W. and Patterson, S.C. (1977) Reform of Congressional campaign spending. *Policy Studies Journal, 5*, 424–31

Crewe, I. (1985a) Great Britain. In I. Crewe and D. Denver (eds.), *Electoral change in Western democracies*, Croom Helm, London, pp. 100–50

Crewe, I. (1985b) MPs and their constituents in Britain: how strong are the links? In V. Bogdanor (ed.), *Representatives of the people?*, Gower Press, Aldershot, pp. 44–65

Crewe, I. and Denver, D. (eds) (1985) *Electoral change in Western democracies*. Croom Helm, London

Crewe, I. and Fox, A. (1984) *British parliamentary constituencies: a statistical compendium*. Faber and Faber, London

Crewe, I. and Payne, C. (1976) Another game with nature: an ecological regression model of the British two-party vote ratio in 1970. *British Journal of Political Science, 6*, 43–81

Curtice, J. and Steed, M. (1980) The results analysed. In D. Butler and D. Kavanagh, *The British General Election of 1979*, Macmillan, London

Curtice, J. and Steed, M. (1982) Electoral choice and the production of government. *British Journal of Political Science, 12*, 249–98

Curtice, J. and Steed, M. (1983) Turning dreams into reality: the division of constituencies between the Liberals and the Social Democrats. *Parliamentary Affairs, 36*, 166–82

Curtice, J. and Steed, M. (1984) The results analysed. In D. Butler and D. Kavanagh, *The British General Election of 1983*, Macmillan, London, pp. 333–73

Dunleavy, P. (1979) The urban basis of political alignment. *British Journal of Political Science, 9*, 409–43

Dunleavy, P. and Husbands, C.T. (1985) *British democracy at the crossroads*. George Allen and Unwin, London

Franklin, M.N. (1985) *The decline of class voting in Britain*. Oxford University Press, Oxford

210

Giddens, A. (1981) *A contemporary critique of historical materialism.* Macmillan, London

Giddens, A. (1984) *The constitution of society.* Polity Press, Cambridge

Goldenberg, E.N. and Traugott, M.W. (1985) *Campaigning for Congress.* Congressional Quarterly Inc., Washington, DC

Gordon, I. and Whiteley, P. (1980) Comment: Johnston on campaign expenditure and the efficacy of advertising. *Political Studies, 28*, 293–4

Hampton, W. (1920) *Democracy and community.* Oxford University Press, Oxford

Hansard Society Commission (1981) *Paying for politics.* The Hansard Society, London

Hay, A.M. (1985) Statistical tests in the absence of samples. *Professional Geographer, 37*, 334–8

Heath, A., Jowell, R. and Curtice, J. (1985) *How Britain votes.* Pergamon, Oxford

Hill, A.P. (1974) The effect of party organisation: election expenses and the 1970 election. *Political Studies, 22*

Himmelweit, H. *et al.* (1985) *How voters decide.* Open University Press, Milton Keynes

Home Office (1984) *European Assembly Election expenses, United Kingdom, June 1984.* Home Office Statistical Bulletin 26/84, Surbiton, Surrey

Houghton Report (1976) *Report of the Committee on Financial Aid to Political Parties.* Cmnd 6601, HMSO, London

Isenberg, S. (1980) Can you spend your way into the House of Commons? *Optimum, 1*, 28–39

Jacobson, G.C. (1978) The effects of campaign spending in Congressional elections. *American Political Science Review, 72*, 469–91

Jacobson, G.C. (1980) *Money in congressional elections.* Yale University Press, New Haven

Jacobson, G.C. (1983) *The politics of congressional elections:* Little Brown, Toronto

Jacobson, G.C. (1984) Money in the 1980 and 1982 congressional elections. In M.J. Malbin (ed.), *Money and politics in the United States,* American Enterprise Institute, Chatham, NJ, pp. 38–69

Jacobson, G.C. (1985) Money and votes reconsidered: congressional elections 1972–1982. *Public Choice, 47*, 7–62

Johnston, R.J. (1977) The electoral geography of an election campaign. *Scottish Geographical Magazine, 93*, 98–108

Johnston, R.J. (1979a) Campaign expenditure and the efficacy of advertising at the 1974 general election in England. *Political Studies, 27*, 114–19

Johnston, R.J. (1979b) Campaign spending and votes: a reconsideration. *Public Choice, 33*, 97–106

Johnston, R.J. (1980a) *The geography of federal spending in the United States of America.* John Wiley, Chichester

Johnston, R.J. (1980b) Electoral geography and political geography. *Australian Geographical Studies, 18*, 37–50

Johnston, R.J. (1983a) Spatial continuity and individual variability. *Electoral Studies, 2*, 53–68

Johnston, R.J. (1983b) The feedback component of the pork barrel: tests using results of the 1983 general election in Britain. *Environment and Planning A, 15,* 1507–16

Johnston, R.J. (1983c) Campaign spending and voting in England: analyses of the efficacy of political advertising. *Environment and Planning C: Government and Policy, 1,* 117–26

Johnston, R.J. (1984) The world is our oyster. *Transactions, Institute of British Geographers, NS9,* 443–59

Johnston, R.J. (1985a) *The geography of English politics: the 1983 general election.* Croom Helm, London

Johnston, R.J. (1985b) A note on local spending in the 1983 general election: differences between the Liberal and SDP parties in England. *Environment and Planning A, 17,* 1393–400

Johnston, R.J. (1985c) People, places and parliaments: a geographical perspective on electoral reform in Great Britain. *Geographical Journal, 151,* 327–38

Johnston, R.J. (1985d) Party strength, incumbency and campaign spending as influences on voting in four English general elections. *Tijdschrift voor Economische en Sociale Geografie, 86,* 82–6

Johnston, R.J. (1985e) Political advertising and the geography of voting in England at the 1983 general election. *International Journal of Advertising, 4,* 1–10

Johnston, R.J. (1985f) Places matter. *Irish Geography, 18,* 58–63

Johnston, R.J. (1986a) The neighbourhood effect revisited: spatial science or political regionalism. *Environment and Planning D: Society and Space, 4,* 41–55

Johnston, R.J. (1986b) A space for place (or a place for space) in British psephology. *Environment and Planning A, 19,* 599–618

Johnston, R.J. (1986c) Placing politics. *Political Geography Quarterly, 5,* S63–S78

Johnston, R.J. (1986d) *On human geography.* Basil Blackwell, Oxford

Johnston, R.J. (1986e) A further look at British political finance. *Political Studies, 34,* 466–73

Johnston, R.J. (1986f) Information flows and votes: an analysis of local campaign spending in England, 1983. *Geoforum, 17,* 69–79

Johnston, R.J. (1986g) Places, campaigns and votes. *Political Geography Quarterly, 5,* S105–S118

Johnston, R.J. (1986h) Information provision and individual behavior: a case study of voting at an English general election. *Geographical Analysis, 18,* 129–41

Johnston, R.J. (1987a) Dealignment, volatility and electoral geography. *Comparative International Development*

Johnston, R.J. (1987b) The geography of the working class and the geography of the Labour vote in England, 1983. *Political Geography Quarterly, 6,* 7–16

Johnston, R.J. and Hay, A.M. (1983) Voter transition probability estimates: an entropy-maximising approach. *European Journal of Political Research, 11,* 93–8

Johnston, R.J. and Taylor, P.J. (1986) Political geography: a politics of places within places. *Parliamentary Affairs, 39,* 135–49

Johnston, R.J. and Wrigley, N. (1988) *Multivariate statistical analysis in geography*. Longman, London

Johnston, R.J., O'Neill, A.B. and Taylor, P.J. (1986) The geography of party support: comparative studies in electoral stability. In M.J. Holler (ed.), *The logic of multi-party systems*, Physica-Verlag, Vienna

Jones, K. (1984) Geographical methods for exploring relationships. In G. Bahrenberg, M.M. Fischer and P. Nijkamp (eds), *Recent developments in spatial data analysis*, Gower, Aldershot, pp. 215–30

Kavanagh, D. (1970) *Constituency electioneering in Britain*. Longman, London

Kavanagh, D. (1986) How we vote now. *Electoral Studies, 5*, 19–28

McAllister, I. (1987) Social context, turnout and the vote: Australian and British comparisons. *Political Geography Quarterly, 7*, 17–30

McCallum, R.B. and Readman, A. (1947) *The British general election of 1945*. Oxford University Press, Oxford

Massey, D. (984) *Spatial divisions of labour*. Macmillan, London

Miller, W.L. (1977) *Electoral dynamics*. Macmillan, London

Miller, W.L. (1984) There was no alternative: the British general election of 1983. *Parliamentary Affairs, 32*, 376–82

Mughan, A. (1986) *Party and participation in British elections*. Frances Pinter, London

Nicholas, H.G. (1951) *The British general election of 1950*. Macmillan, London

Norton, P. (1978) *Conservative dissidents*. Temple Smith, London

O'Leary, C. (1962) *The elimination of corrupt practices in British elections 1868–1911*. Clarendon Press, Oxford

Owens, J.R. and Olson, E.C. (1977) Campaign spending and the electoral process in California, 1966–1974. *Western Political Quarterly, 30*, 493–511

Palda, K.S. (1973) Does advertising influence votes? An analysis of the 1966 and 1970 Quebec elections. *Canadian Journal of Political Science, 6*, 638–55

Palda, K.S. (1975) The effect of expenditures on political success. *Journal of Law and Economics, 18*, 745–71

Paltiel, K.Z. (1981) Campaign finance: contrasting practices and reforms. In D. Butler, H.R. Penniman and A. Ranney (eds), *Democracy at the polls*, American Enterprise Institute, Washington, DC, pp. 138–72

Patterson, S.C. (1982) Campaign spending in contests for governor. *Western Political Quarterly, 35*, 457–77

Patterson, S. and Caldeira, G.A. (1983) Getting out the vote: participation in gubernatorial elections. *American Political Science Review, 77*, 675–89

Pennock, J.K. (1932) *Money and politics abroad*. A.A. Knopf, New York

Pinto-Duschinsky, M. (1981a) *British political finance 1830–1980*. American Enterprise Institute, Washington, DC

Pinto-Duschinsky, M. (1981b) Financing the British general election of 1979. In H.R. Penniman (ed.), *Britain at the polls 1979*, American Enterprise Institute, Washington, DC, pp. 210–40

Pinto-Duschinsky, M. (1985) Trends in British political funding. *Parliamentary Affairs, 38*, 328–47

Przeworski, A. (1985) *Capitalism and social democracy*. Cambridge University Press, Cambridge

Rose, R. (1961) Money and election law. *Political Studies, 9*, 1–15

Rose, R. (1976) *The problem of party government*. Penguin, London

Rose, R. and McAllister, I. (1986) *Voters begin to choose*. Sage Publications, London

Sarlvik, B. and Crewe, I. (1983) *Decade of dealignment*. Cambridge University Press, Cambridge

Savage, M. (1987) Understanding political alignments in contemporary Britain: do localities matter? *Political Geography Quarterly, 6*, 53–76

Scarbrough, E. (1984) *Political ideology and voting*. Clarendon Press, Oxford

Seymour, C. (1950) *Electoral reform in England and Wales* (first printed in 1913). David and Charles, Newton Abbot

Taylor, A.H. (1972) The effect of party organization: correlation between campaign expenditure and voting in the 1970 election. *Political Studies, 20*, 329–31

Taylor, P.J. (1978) Political geography. *Progress in Human Geography, 2*, 153–62

Taylor, P.J. (1979) The changing geography of representation in Britain. *Area, 11*, 289–94

Taylor, P.J. (1982) The changing political map. In R.J. Johnston and J.C. Doornkamp (eds), *The changing geography of the United Kingdom*, Methuen, London, pp. 275–90

Taylor, P.J. (1985a) *Political geography: world-economy, nation-state and locality*. Longman, London

Taylor, P.J. (1985b) The geography of elections. In M. Pacione (ed.), *Progress in political geography*, Croom Helm, London, pp. 243–72

Taylor, P.J. (1986) An exploration into world-systems analysis of political parties. *Political Geography Quarterly, 5*, S5–S20

Taylor, P.J. and Johnston, R.J. (1979) *Geography of elections*. Penguin, London

Thrift, N.J. (1983) On the determination of social action in space and time. *Environment and Planning D: Society and Space, 1*, 23–57

Todd, D. (1980) *An introduction to the use of simultaneous equation regression analysis in geography*. CATMOG 21, Geo Books, Norwich

Tuckel, P.S. and Tejera, F. (1983) Changing patterns in American voting behavior, 1914–1980. *Public Opinion Quarterly, 47*, 230–46

Waller, R.J. (1984) *The almanac of British politics*. Croom Helm, London

Warde, A. (1986) Space, class and voting in Britain. In K. Hoggart and E. Kofman (eds), *Politics, geography and social stratification*, Croom Helm, London, pp. 33–60

Welch, W.P. (1974) The economics of campaign funds. *Public Choice, 20*, 83–99

Welch, W.P. (1976) The effectiveness of expenditure in State legislature races. *American Politics Quarterly, 4*, 333–56

Welch, W.P. (1981) Money and votes: a simultaneous equation model. *Public Choice, 36*, 209–34

Welch, W.P. (1982) Campaign-contributions and legislative voting: milk money and dairy price supports. *Western Political Quarterly, 35*, 479–95

Whiteley, P. (1983) *The Labour Party in crisis*. Methuen, London

Whiteley, P. (1986) Predicting the Labour vote in 1983: social backgrounds versus subjective evaluations. *Political Studies, 34*, 82–98

Wrigley, N. (1984) Quantitative methods: diagnostics revisited. *Progress in Human Geography, 8*, 525–35

Index

Printed and bound by CPI Group (UK) Ltd, Croydon, CR0 4YY

22/10/2024

01777621-0009